ANIMATING CULTURE

Hollywood Cartoons from the Sound Era

ERIC SMOODIN

RUTGERS UNIVERSITY PRESS New Brunswick, New Jersey

Library of Congress Cataloging-in-Publication Data

Smoodin, Eric Loren.
Animating culture : Hollywood cartoons from the sound era / Eric Smoodin.
p. cm. — (The Rutgers series in communications, media, and culture)
Includes bibliographical references and index.
ISBN 0-8135-1948-9 (cloth) — ISBN 0-8135-1949-7 (pbk.)
1. Animated films—United States—History. 2. Animated films—Political aspects—United States. I. Title. II. Series.
NC1766.U5S66 1993
791.43'3—dc20 92-32891
CIP

Manufactured in the United States of America

For Caren

Certainly. Why Not?

CONTENTS

ILLUSTRATIONS

PREFACE

Watching upwards of four hundred cartoons over four years can alter anybody's way of thinking. My own conception of this project changed well into my steady viewing of animated shorts, which began in 1987. Animation constituted something of an alternative production practice within the Hollywood studio system, and I believed that, to understand the latter more fully, we would have to analyze the former (and also, in time, newsreels, coming attractions, short subjects, and other non–feature-length Hollywood products). When I first began working on this project, I planned an exhaustive study of the cartoon narrative, visual strategy, genres, star system and so on. I imagined using the Hollywood cartoon itself as the primary location of meaning, as the text that could be used to explain a number of contexts.

One of the strengths of such an approach is that, except for a few Walt Disney films, there are no acknowledged masterworks of Hollywood animation. All of us are familiar with, and probably fond of, Betty Boop and Porky Pig, but few of us can name any of their titles. Studying animation means studying decidedly minor films, which nonetheless entertained hundreds of millions of people. Cartoons, then, far more than any live-action masterpieces, explain the workings of a Hollywood machine that turned out products that could be matchlessly entertaining and instantly forgettable at the same time.

I soon became less interested in issues related primarily to textual analysis, however, and more concerned with those practices that themselves help us to understand cartoons and the political, cultural, and industrial network of which they were a part. My original methodology became a multi-methodology, shifting from a concentration on the films to an emphasis on not only cartoons, but also copyright records, censorship codes, popular journalism, government documents, and the interaction of cartoons with other types of movies.

Shifting the emphasis to a range of related practices helps to eliminate one of the problems that has dogged film studies, even as historians and critics have moved away from a consideration of masterpieces and toward an interest in movies that have always been considered run-of-the-mill. When we study huge bodies of film, unless we can generate an unbiased sample, and then have access to the films in that sample, we can never be sure, regardless of how many films we watch, that we have isolated a representative group.[1]

The problem only worsens when cartoons become the object of study. Long marginalized as children's entertainment, cartoons have rarely been collected by film archives. Because of this, films for minor cartoon studios, and even those major studio cartoons that lack popular characters (Warner Bros. animated shorts from the early 1930s, for instance), are almost impossible to see. Moreover, black-and-white cartoons were colorized long before the issue caused a furor among those who could not bear the thought of a pastel *Maltese Falcon*. Cartoons have also been tampered with for television far more ruthlessly than most feature films (in just one example, the stereotypical black maid in several *Tom and Jerry* cartoons became, for 1950s and 1960's TV audiences, stereotypically Irish). Thus, the student of animation is limited to a cartoon canon formed by those films released on videotape and that play routinely on television, many of which may be available only in adulterated versions.

In general, studies that center primarily on the films themselves also pose difficulties in our understanding of movie reception. A virtuoso reading of a film or group of films frequently begs the question, "Who is this reading available to?" and tends to make movies impervious to the different experiences of different members of the audience. One of the significant contributions of poststructuralist and psychoanalytic approaches to film studies has been their interest in the spectator, and more specifically in the interactive relationship between spectator and text. Recently, studies seeking to theorize film history have expanded this field of inquiry, refining our understanding of the various groups that make up any film audience and showing how the meaning produced by film and spectator comes to be created by an institutional network made up of advertising, radio shows, star interviews, censorship codes, and so on.

Lea Jacobs, George Custen, and Jane Gaines have been instrumental in adapting this type of analysis to film and cultural studies,[2] and it is a method that I have adopted here, largely because of my own experiences in the classroom. I have found my students only too willing to consider American society in monolithic terms, and to see films simply as working to reflect and heighten unanimity. They will read *Casablanca*, for instance, as indicative of the overwhelming support for World War II among United States citizens, and also of the movie industry's attempt to make sure that that support would never waver. Or a *Rambo* movie both demonstrates and fosters an unquestioned loathing of all things Soviet during the 1980s.

As a result, in my classes and also in this book, I use a variety of related sources to show that this kind of film analysis, though not wrong, is always incomplete. Different materials dealing with similar issues show us possibilities for a number of readings of any film by a number of different audiences, both at the time of the film's release and long after it. Rather than explaining Hollywood animation, then, I hope that this book, by concentrating on cartoons, demonstrates the manner in which meaning is always contested and variable.

Although I believe them to be, I cannot claim that the cartoons I discuss here are absolutely representative of Hollywood animation production. Neither can I assert that the FBI files that I cite, or the government documents, or the copyright records, are representative of all such papers. Reading all of these films and documents together, however, and analyzing the concerns that they share reveal many of the most significant tensions that informed not only the production of Hollywood movies, but also the experience of viewing them.

ACKNOWLEDGMENTS

As difficult as it is to know for sure when a project has been finished, it is often just as hard to determine precisely when it began. While I was writing this book, my mother, Mildred Smoodin, provided me with all manner of support. But I would also like to thank her and my father, Solly Smoodin, for creating an atmosphere when I was growing up in which watching movies all night and then talking about them could be both great fun and serious business. Many years ago my sister, Roberta Smoodin, gave me my first and best writing lessons, and by example and through advice continues to teach me what it means to be a writer.

Doris, Arthur, and Mitchell Kaplan have been constant sources of good sense, professional expertise, and kindness. During a difficult period, Steve Marvin helped me to accentuate the positive.

David Lash was a source of excellent conversation, legal assistance, and emotional support, all equally appreciated. His entire family took an active interest in my work, and were quick to help in any way they could.

I also benefited from the generosity, good judgment, and good company of both Henry Flax and Mark Zakarin. Mark Anderson unselfishly shared ideas that helped me sharpen my arguments. Nick Browne and Thomas Elsaesser, two graduate school mentors who have maintained an interest in my career and my work, provided encouragement and assistance. Richard deCordova, Jon Lewis, Frank Tomasulo, and Richard Neupert, my friends since graduate school, have always been available to discuss issues from the professional to the personal. Friends from long before that provided both long distance and local help: Fred Davidson in particular, but also Robert Ring, Robert Zipser, Dan Landau, Allan Favish, and Russell Stein.

Other friends and colleagues were always willing to advise, consult, and encourage: Katie King, Michael Ragussis, Pamela Fox, Denise Albanese, Gina Marchetti, Inderpal Grewal, Ann Martin, Kim Hall, Chris Grella, Margaret Stetz, Mark Samuels-Lasner, Vivian Sobchack, Bill Pietz, E. Ann Kaplan, Krin Gabbard, Sigmund Roos, Ella Shohat, Susan Ohmer, Leyla Ezdinli, Arnold Grossblatt, Karla McPherson, Lynn Thiesmeyer, Heather Hendershot, and Harvey Grossinger. Still others provided research information and professional support: Dana Polan, Tino Balio, Robert Eberwein, Charles Maland, Rick Altman, and Paul Swann.

Many of the people with whom I have worked at American University were generous with their help and advice: Jo Radner, Deborah Payne, Huma Ibrahim, Julie Solomon, Richard McCann, Dan Crawford, Charles Larson, Jeanne Roberts, Frank Turaj, Henry Taylor, Ron Sutton, and Frank Zapatka. The College of Arts and Sciences at the university supported my research with various grants.

Alan Kraut was both generous and tireless in his support, while John Colwell and Mary O'Farrell provided extremely valuable professional services. At the Library of Congress, Rosemary Hanes patiently answered all of the questions I asked her and gave me research materials that I did not know existed. Rose Motzko was similarly helpful at the Walt Disney Archive in Burbank, California. Staff members at the National Archives, and particularly those overseeing records from government film productions and from the Treasury Department, were unfailingly friendly and always eager to assist in my research. Elisa Komins lent her expertise as a photographer, and Carl Montuori and Harry Zalewski at Video Lab in Rockville, Maryland, supplied technological help.

My editors, Leslie Mitchner, George Custen, and Marilyn Campbell, have been rigorous and kind. This is a much better book because of their efforts.

These people and others helped make this project possible. But more than anyone else, I want to thank Caren Kaplan. She has been my best colleague, collaborator, teacher, and friend. And with good humor and great patience she has put up with numberless Looney Tunes, Merrie Melodies, and Silly Symphonies.

ANIMATING CULTURE

INTRODUCTION: HOLLYWOOD ANIMATION AND SOCIAL CONTROL

Hollywood animation has received little scholarly attention, and has routinely been considered by audiences and the popular media, particularly during the last forty years, as children's entertainment.[1] But theatrical cartoons in general and Walt Disney's in particular have been implicated in a number of discourses about power, behavior, and social control. A 1948 article in *The Saturday Review of Literature* warned that in postwar cartoons, the once cute and fuzzy animals had become "masks for human characters so violent and crude that they couldn't even be put on the screen in their natural form."[2] Then, in 1959, FBI Director J. Edgar Hoover alerted a presidential assistant about Walt Disney's nomination to the Advisory Committee on the Arts. Hoover acknowledged that Disney had never been investigated by the FBI, but warned that he had appeared in 1944 at a tribute to the late cartoonist Art Young, sponsored by a newspaper called *New Masses*. "According to the Special Committee on Un-American Activities," Hoover cautioned, the *New Masses* was a "nationally circulated weekly journal of the Communist Party."[3] Evidently, the cartoons from the sound era of Hollywood's classical period (roughly 1928 to 1960)[4] and that period's leading cartoon entrepreneur were frequently perceived by cultural arbiters and the government to wield considerable power.

Within the film industry itself, however, cartoons suffered in comparison with the dominant Hollywood product, the feature-length, live-action fiction film. Exhibition practices relegated cartoons to less important positions on the film bill than those held by the feature film or the newsreel, while those feature-length, live-action films either ignored cartoons altogether or contained them within the interstices of the movie narrative. In *Sabotage* (1936), for instance, we see and hear brief scenes from Disney's *Who Killed Cock Robin?* (1935), the cartoon that serves as a backdrop for a movie theater chase scene. Toward the end of *Bringing Up Baby* (1937), a wrongly-incarcerated Cary Grant talks about his partners in crime, Mickey the Mouse and Donald the Duck, while the police dutifully take down the names. In *Having Wonderful Time* (1938), Douglas Fairbanks, Jr., refuses to go into a party because the guests periodically break into choruses of the *Snow White* anthem, "Heigh Ho." Both star and camera stay outside, with the singers barely visible, while the song remains audible enough to induce Fairbanks's headache. And in these last two films, as in a number of other movies produced by RKO during the late 1930s, references to Disney animation appeared only because RKO itself served as Disney's distributor. Once the association with Disney ended, so too did the advertisements for his films.

Despite this poor-cousin status, cartoons were a staple of Hollywood production during the classical period. Leonard Maltin's cartoon studio filmography in *Of Mice and Magic: A History of American Animated Cartoons,* lists some five thousand cartoons, almost all of them of the six-to-eight-minute variety, and this does not include the cartoons produced for use by the military, for example, or in the classroom.[5] Throughout the period under consideration, the same studios that dominated feature-film production also controlled cartoon production. Some studios released cartoons produced by independent animation studios—for several years in the 1930s, for instance, RKO handled not only all of the Disney product, but also the cartoons from the less well known Van Beuren Corporation, while Universal distributed the Walter Lantz cartoons (which included, among other series, "Woody Woodpecker"). Other studios, like MGM, had their own animation units. As with feature films, cartoon production decreased throughout the period, but animation always accounted for a significant proportion of a studio's product. For example, Warner Bros. released forty "Looney Tunes" and "Merrie Melodies" in 1940, thirty-one in 1950, and twenty in 1960 (in comparison, Warner Bros. released forty-six feature-length films in 1940, and twenty-three in 1949, the last year for which I could find precise information).[6]

Analyzing cartoons and the practices related to them complicates our understanding of film reception, and of the uses made of movies by such institutions as the government, the law, and journalism. Cartoon copyright records, for instance, can show us the possibilities for different readings of the same film. Or, the arrangements of movies on the film bill, or the information about cartoons provided by popular journalism, probably affected the manner in which audiences understood all manner of movies. As a result, my concern will be not so much with producing an encyclopedic guide to such codes of intelligibility as editing, lighting, color, or genre, although these codes, in relation to animation, are of course worthy of detailed study. I am instead more interested in the intelligibility of cartoons in relation to audience, film exhibition, newspapers and magazines, censorship, and government.

Animation production participated with other Hollywood products in a system of regulation and control. Chapter 1 takes as its starting point the regulation of cartoons themselves through industry self-censorship. While cartoon sexuality was controlled by Hollywood's Production Code, animation also stretched the code or openly battled with it, in part for reasons related to studio competition, audience demographics, and historical context. Moreover, cartoons from the classical period and the Production Code itself were simply two aspects of a regulatory practice that included copyright law, articles in popular journals, and the social sciences, with the strains between them complicating our notions of audience and interpretation.

Pierre Bourdieu has written of the necessity "to establish the conditions in which the consumers of cultural goods, and their taste for them, are produced,"[7] and this is precisely the project of Chapters 2 and 3 in relation to cartoon viewing. The cartoon's place on the first-run film bill, and the entire bill itself, demonstrate the very careful construction of an afternoon's or evening's entertainment. At once related to the vaudeville experience and the insistence, dating at least from the nineteenth century, on regulating time and space, the bill also typified the cinema as a "technology of simultaneity," to use Stephen Kern's phrase.[8] That is, the combination of entertainments—the feature film, the newsreel, the travelogue, the cartoon, and the live show—allowed not only for all genres, but also for all cultures from all times to coexist; or, at least, Hollywood constructions of all cultures from all times. A historical drama might play with a newsreel of contemporary, nonfiction events, a short documentary about Africa, and a fairy-tale cartoon.

In addition to signifying Hollywood universality and timelessness, however, the combination of films on the bill also fulfilled the function of justifying social differences. Indeed, first-run theaters used the film bill to provide the world to the middle class, and created a space for entertainment that, either through demographics or legal restrictions (segregation laws, for instance), provided a safe haven for that class.

The film bill also constructed, as entertainment, Hollywood monopoly practice. By so signifying entertainment through the variety of films, those few studios that controlled film production and first-run exhibition (they owned many of the largest urban theaters) made other companies and other exhibition venues seem unnecessary. The entrance of the United States into World War II saw an extension of this exhibition strategy along with slight changes in it, and under the control of a different studio, the United States government, which nevertheless worked in close cooperation with Hollywood.

Starting in 1943, Frank Capra oversaw the twice-monthly *Army-Navy Screen Magazine,* a twenty-minute selection of films shown primarily to servicemen, which often included a Warner Bros.-produced "Private Snafu" cartoon. As opposed to the first-run theatrical bills, which played primarily to a middle-class audience and thus worked to assert that class's special social status, the films in the *Army-Navy Screen Magazine* played to Americans from different class backgrounds, different levels of educational attainment, and differing tolerance to military authority. Because of this, rather than legitimating social differences, the "Private Snafu" cartoons and the other films in the *Magazine* sought to make them disappear, and to replace them with the appearance of ideological consensus about the war.

In both cases, that of the domestic film bill and the military variety, we see the emergence of a kind of cultural production that could impose its own norms and regulations on both the product and its consumption. The combination of major studio and United States military in the production of the *Magazine* demonstrates the confluence of the corporate and the governmental during the period, and indicates the similar goal of each, namely promoting the belief on the part of the spectator in that which was represented.

The last two chapters take a more careful look at the means of constructing that belief, and also at the possibilities for resisting it. Walt Disney dominated both the public discourse about animation and the private. If the combination of films on the theatrical bill helped to construct the film audience, then the audience, too, served to construct the films themselves. Mainstream magazines like *Popular Science* and *The Saturday Review* functioned prominently at this point of exchange between film viewer and film text, explaining how to interpret the Disney product, not only aesthetically but also ideologically. In so treating Disney as a subject worthy of multifaceted discussion, the journals also disseminated instructions for the proper attitudes toward such issues as gender, race, class, and labor.

The Walt Disney thereby created varied only slightly from journal to journal, from *Time* magazine, for instance, to *Business Week*. But the private discourse about Disney, produced by the United States government, demonstrated all of the contradictions inherent in the government's project of promoting capitalism and democracy at home and abroad. State Department documents detail Disney's work for the U.S. government in South America during the 1940s, as a representative of the Good Neighbor policy and of North American industry. But it was largely this progovernment work in South America and the increase in the global influence of the Disney product that made the FBI cautiously suspicious of the cartoon producer. Disney typically is depicted as a purveyor of dominant values, as in Richard Schickel's 1968 study, *The Disney Version*.[9] But the government documents show that he and his product formed something of a contested zone, which various branches of the government attempted to manipulate for their own ends.

Finally, Treasury Department documents make us question the very dominance of those values that we have come to associate with Disney. The responses to his 1942 cartoon *The New Spirit* indicate that, at least early in the U.S. war effort, Disney himself did not function as an unproblematic icon of entertainment and education. Further, in their favorable and unfavorable attitudes about the film, the viewing public in 1942 showed how deeply divided they were along class lines and in relation to such issues as race and patriotism, and how resistant they could be to a wartime ideology of consensus.

Notions of resistance inform the entire book. By this, however, I do not simply mean Third World resistance against the First, or the possibility for even a small domestic audience to resist the representational strategies of dominant ideology. Instead, I also mean resistance within that seemingly monolithic network that constructs dominant ideology: resistance to censorship within the industry, for instance, or Disney's own problematic relationship with the government (his refusal, for example, to accede to the FBI's insistence on reviewing the script for *That Darn Cat*). My concern, then, is with the exercise of power in any number of relationships, and with the effectiveness of that power and the possibility for both its spread and its containment. This project carries on from such recent studies as Dana Polan's of the Hollywood cinema of the 1940s, in which Polan seeks to understand both the various representations of power, and the manner in which those representations foster that power and simultaneously construct its limits.[10] At least implicitly, the chapters that follow work to support the concerns of J. Edgar Hoover and *The Saturday Review,* that even cartoons, those childish and unimportant cultural artifacts, are intertwined with behavior, with social control, and even with transnational relations.

1 | STUDIO STRATEGIES

Sexuality, the Law, and Corporate Competition

From the beginnings of the cinema as a commercial enterprise at the end of the nineteenth century, the moral content of films, and in particular the cinema's representation of sexuality, was an area of special concern for educators, social reformers, religious groups, and social scientists.[1] Throughout the silent period, the film industry sought to balance realistic marketing strategies ("sex sells") against the problems of city, state, and even federal regulatory practice. Individual states passed their own film censorship codes, and the federal government also possessed the potential for sweeping police action in regard to film content, a potential made all the more real by the government's willingness to regulate certain industries in the national interest, such as railroads, during World War I.

Among the various forms of Hollywood film product from the classical period, cartoons can be used to analyze censorship because of the manner in which they, and the various practices that dealt with them, demonstrate a particularly modern concern with sex. In the United States, the last two centuries have been marked by a separation of adult and child in most social settings. The theatrical film bill itself from Hollywood's classical period marks one of the fullest developments of this phenomenon and also one of its contradictions. To maximize profits, theaters had to attract, on the one hand, as wide an audience as possible. On the other, they had to appeal to the various segments of that audience. Although these distinctions were not exact during much of the period under discussion, the feature film and the newsreel might have appealed mostly to adults, while the serial or cartoon might have catered to the tastes of adolescents. That which was said about cartoons during the 1930s and

'40s, then (before cartoons came to be more rigidly fixed as children's entertainment in the 1950s) was also being said about young people, about their perceived needs, and about the perceived dangers that they faced.

As part of its continuing project to avoid federal regulation, and also to demonstrate concern with the moral education of children and young adults, the Hollywood cinema adopted a self-censoring Production Code in 1930. A few years later, however, a number of citizens groups, most notably the newly formed Catholic Legion of Decency, still were unconvinced of Hollywood's sincerity about improving the moral content of its product. By threatening audience boycotts and other hostile actions, the Legion prodded the movie studios to adhere more strictly to the Code.

The Code addressed the issue of the diverse audience for motion pictures and the popularity of movies among children. An epilogue to the Code, explaining the necessity for it, stresses that while "most arts appeal to the mature," movies attract both the "mature and immature," and then, just two paragraphs later, employs the same phrase about broad appeal once again.[2] To protect "mature and immature" alike, the Code concerned itself particularly with representations of sexuality, and especially with regulating depictions of what was considered excessive female sexuality. In the Code's section on sex, the nine caveats—against "lustful" kissing or embraces, or against passion that might "stimulate the lower and baser emotions"—apparently applied equally to men and women (although there was also a caution against white slavery). In other parts of the Code, however, female sexuality became the focus of censorship. The conditions regulating costuming insisted on no nudity, no undressing, and no "indecent nor undue exposure," all of which alluded to the occasional scenes of nude women that one finds in films from the silent and early sound periods. Additionally, in the category of profanity, most of the gender-specific terms refer to women, and most of those indicate a certain type of sexual behavior: "alley cat," "chippie," "cocotte," "tart," "madam," "whore," and "slut," for instance. Only a few terms, such as "tom cat," refer to male sexual behavior, and most of those—"fairy," for instance, and "pansy"—refer to male homosexuality, which itself during the classical period often came to be represented as feminized male behavior. Further, as some recent historians have shown, several of the most pitched battles between the Hays Office, which administered the Code, and the Hollywood studios were fought precisely over degrees of female sexuality: whether or not Marlene Dietrich, in

Blonde Venus (1932), could be forgiven for her adultery, or whether Mae West, in *Belle of the Nineties* (1934), would be allowed her double-entendre–laden dialogue.[3]

CENSORSHIP, SOCIAL SCIENCE, AND COPYRIGHT

The Production Code concerned itself primarily with the regulation of live-action, feature-length motion pictures. But it also applied to all manner of Hollywood productions, including animation. Indeed, in 1939, *Look* magazine published an article with pictures called "Hollywood Censors Its Animated Cartoons" that demonstrated the studios' concern with cartoon content.[4] Like the Production Code itself, the piece indicates something of the institutional network in which Hollywood animation operated. The article illustrates the period's conviction that controlling film content was a means of regulating audience desire and behavior. Moreover, while not necessarily doubting Hollywood's commitment to moral uplift, one can also read the *Look* article, as I will later in this chapter, as a document that, while talking about censorship, actually addresses many of the commercial concerns of all of the animation studios, or, more properly, all of the animation studios except Walt Disney's.

Anecdotally and with amusing pictures, articles like those in *Look* disseminated censorship information to a wide audience, frequently to certify the movie studios' efforts to provide wholesome entertainment. During the same period, and indicating the era's faith in the social sciences, the educational, sociological, and psychological communities undertook a major project to measure the effects of the movies upon their audiences, and especially upon the adolescent audience. In particular, the era's experts analyzed the effects of depictions of crime, race relations, and sexuality. From 1929 to 1933, the Payne Fund sponsored this research, and between 1933 and 1935, the years marking the rise of the Catholic Legion of Decency and the strict implementation of the Production Code, the Macmillan Company published a number of the Payne Fund studies charting the interaction of movie and movie audience: for example, *Movies and Conduct, Motion Pictures and Standards of Morality, Getting Ideas from the Movies, Movies, Delinquency, and Crime,* and, in joint editions, *The Social Conduct and Attitudes of Movie Fans* with *Motion Pictures and the Social Attitudes of Children,* and *Children's Attendance at Motion Pictures* with *The Emotional Responses of Children to the Motion Picture Situation.*[5] As Lea Jacobs has pointed out in her groundbreaking work on 1930s audiences, the decade also witnessed the beginning of the Film Education Movement, through which various organizations assembled film appreciation programs for showings in public schools in order to control adolescent response to movies.[6]

Copyright records form a heretofore unexplored source for examining some of the conflicts studios may have had in conforming to censorship codes and also some of the methods adopted by the studios to control objectionable material. Until the early 1940s, motion picture studios did not submit permanent copies of their films to the Library of Congress as part of the copyright process (and even then, the Library requested that only selected films be deposited). Instead, when the studios registered their product at the Library's Office of Copyright, they submitted a copy of the film which, except in special cases, the Library would return. Then, as part of the permanent record of a feature film being copyrighted, studios deposited scripts, publicity material, and other general information. For cartoons, studios usually submitted far less than that: detailed, one- or two-page descriptions in the case of Walt Disney products; transcriptions of dialogue for Paramount cartoons; telegraphic, five- or six-line descriptions for Warner Bros. animated shorts. In 1942, Librarian of Congress Archibald MacLeish instituted a program in which all films submitted for copyright (about 1,400 a year) were analyzed by a small staff whose job it was to determine which movies ought to be preserved in the Library's limited storage facility. The Library hoped for a film collection that would "serve the student of history rather than the student of the movie art as such, a collection that will illuminate in retrospect the periods which have produced the films."[7] These staff analyses, along with the other materials submitted by the studios, became part of the copyright record.

These copyright documents function as what could be called the legal representations of the cartoons, just as the combinations of image and sound form the cartoons themselves. Reading this copyright material alongside the cartoons that they codify reveals the disjunctions between legal discourse and film practice, particularly those disjunctions that emerge in the representations of sexuality. Indeed, those cartoons that seem most constrained by the demands of the Production Code are often those that highlight sex in the copyright documents, while those that seem to demonstrate the least concern with the Hays Office tend to use their written descriptions to proclaim the innocence of the artifact. At stake here in the reading of these various films and documents is not only our understanding of the different and often contradictory aspects of the Hollywood film industry, but our understanding, as well, of how audiences interpreted the movies that they watched.

Film Censorship and Cartoon Economics: Reading *Look* Magazine

"Hollywood Censors Its Animated Cartoons," the 1939 *Look* article, constitutes, for my purposes, an extraordinary document. The piece immediately stands out as unusual because it has nothing to do with Walt Disney. Chapter 4 will recount how Disney dominated discourse about animation in the popular media throughout the 1940s, a domination that had begun at least ten years before. A look through the *Readers' Guide to Periodical Literature* from the 1930s, for example, shows that magazines published hundreds of articles about animation, but only a handful dealt with cartoon studios other than Disney's.

The *Look* article also stands out because censorship information about cartoons is so much more difficult to locate than similar data about other film products. The Hays Office censorship files recently made available by the Academy of Motion Picture Arts and Sciences Library apparently contain no documents about cartoons, and, assuming that the *Readers' Guide* provides an accurate indication, there are no other articles during this period that deal with cartoons and censorship.

"Hollywood Censors Its Animated Cartoons" indicates that even in the 1930s, a decade of intense debate about film censorship, cartoons were left out of the discussion. "You seldom hear of movie cartoon censorship," the article begins, "except when Europe's jealous dictators, in a rage because their people have been caught laughing, bar such sinister foreigners as Daffy Duck, Porky Pig and Popeye." The article reassured its readers, however, that censors did indeed watch over cartoon content, and that "in fact, these cartoons have censorship problems more complex than those of feature pictures."

Look explained that cartoon producers belonged to and were bound by the "Will H. Hays organization," which administered the Production

Code and which approved scripts as well as finished films. "Unlike feature productions, however, which are censored in the script, as well as after shooting," the article continued, "the animated cartoon is passed upon only when completed," and as a result "cartoon producers must be exceptionally cautious as to restrictions."

The article then invoked Leon Schlesinger, the producer of "Merrie Melodies" and "Looney Tunes" for Warner Bros., who supplied pictures for the censorship piece. Showing his concern for the good of the adolescent audience, Schlesinger said: "We cannot forget that while the cartoon today is excellent entertainment for young and old, it is primarily the favorite motion picture fare of children. Hence, we always must keep their best interests at heart by making our product proper for their impressionable minds." The article concluded by saying that "Schlesinger makes 42 cartoons a year."

Photographs of cartoon characters accompany the article, with the captions explaining the prohibitions against kissing, monsters, cruelty to animals, references to God, bodily functions, and spitting, and also Italy's restrictions against parodies of Fascist authority.

For two photos of Porky and Petunia Pig holding hands and then kissing, the first of the captions/caveats explains that "Robert Taylor may kiss Garbo in a feature picture, but it isn't considered nice for Porky Pig to kiss Petunia Pig in an animated cartoon. Censors prefer romance of the hand-holding type." An overlarge scissors, with "censored!" written on it in large letters, cuts through the photos (fig. 1). The prominence of this warning against kissing—it comes at the head of the article, all of the other captions accompany only one photo, and none of the other pictures has a scissors slicing through it—hints at the significance of the issue of how heterosexuality would be represented and regulated, especially in relation to the adolescent audience.

The very appearance of the article tells us something about the film industry in 1939. Almost certainly a Warner Bros. publicity release, the article implies that the industry was still trying hard to consolidate its hold on a mass audience by insisting on animation's edifying characteristics; there will be nothing, in any cartoon, that parents would be embarrassed to have their children watch. The article also hints at the ambivalent position of cartoons within Hollywood production. By stressing animation as pure, wholesome entertainment, the article implicitly alludes to an alternative discourse about animation, one that stressed its connection to commodities. Also in 1939, for instance, Margaret Thorp published her influential *America at the Movies,* one of the best known of

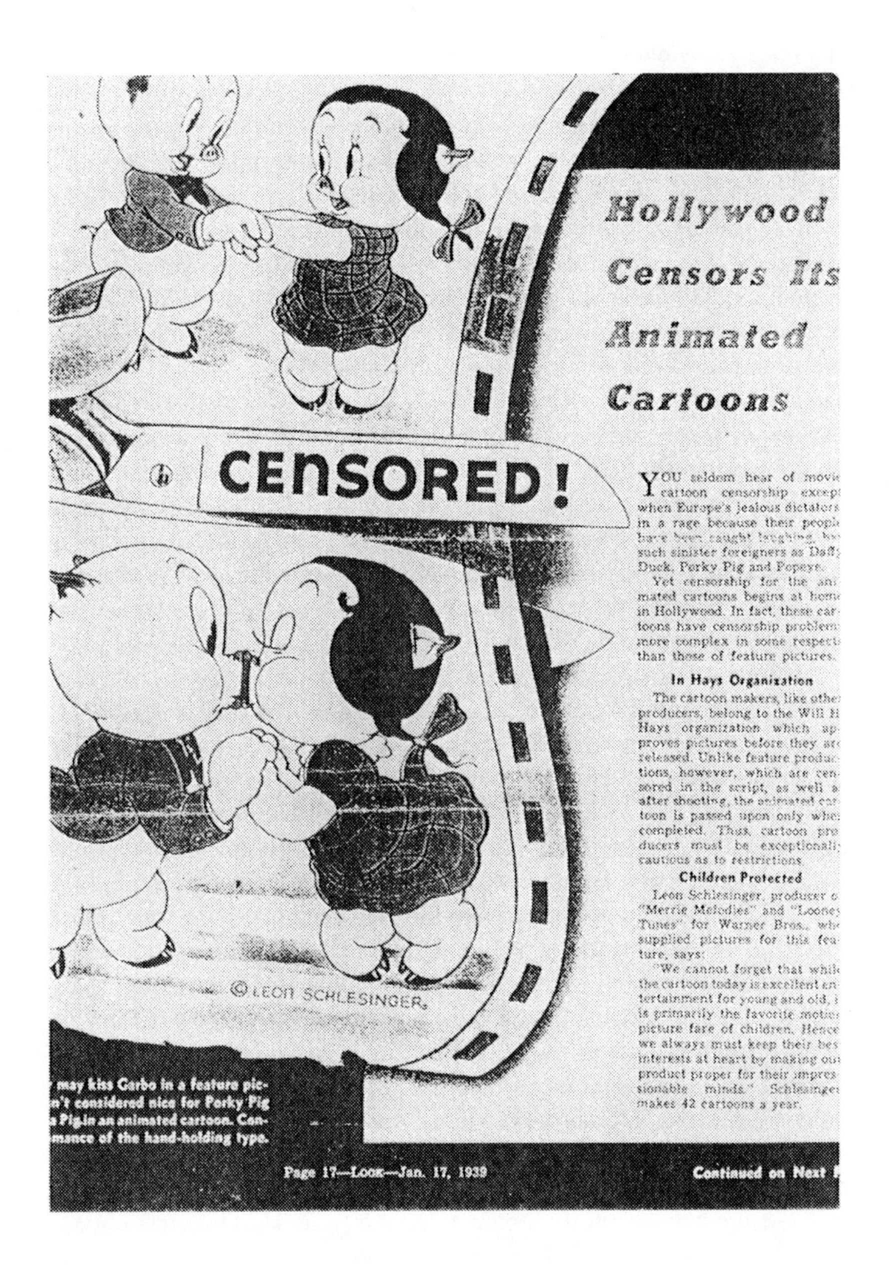

Hollywood Censors Its Animated Cartoons

YOU seldom hear of movi cartoon censorship excep when Europe's jealous dictators in a rage because their peopl have been caught laughing, b such sinister foreigners as Daff Duck, Porky Pig and Popeye.

Yet censorship for the ani mated cartoons begins at hom in Hollywood. In fact, these car toons have censorship problem more complex in some respect than those of feature pictures.

In Hays Organization

The cartoon makers, like othe producers, belong to the Will H Hays organization which ap proves pictures before they ar released. Unlike feature produc tions, however, which are cen sored in the script, as well a after shooting, the animated car toon is passed upon only whe completed. Thus, cartoon pro ducers must be exceptionall cautious as to restrictions.

Children Protected

Leon Schlesinger, producer o "Merrie Melodies" and "Looney Tunes" for Warner Bros., wh supplied pictures for this fea ture, says:

"We cannot forget that whil the cartoon today is excellent en tertainment for young and old, i is primarily the favorite motio picture fare of children. Hence we always must keep their bes interests at heart by making ou product proper for their impres sionable minds." Schlesinge makes 42 cartoons a year.

may kiss Garbo in a feature pic- n't considered nice for Porky Pig s Pig in an animated cartoon. Cen- mance of the hand-holding type.

Page 17—Look—Jan. 17, 1939

Continued on Next

Figure 1. "Hollywood Censors Its Animated Cartoons," *Look* magazine, 17 January 1939.

the era's audience studies. Discussing movie merchandising, Thorp wrote that "the most staggering of all the figures in the merchandising record of the movies are those that number the products made in the likeness of favorite cartoon characters: Mickey Mouse, Donald Duck, the Seven Dwarfs, Popeye." She added that "the less skillful cartoonists have not succeeded in getting their characters into the national culture." Thorp generally lauded these commercial tie-ins (she cites animation as having created "a whole new American folklore," and she further explains that "Teachers College has given academic sanction to this new and exemplary folklore by endorsing a Mickey Mouse primer").[8] But it was also a part of Hollywood's economic strategy to downplay animation production as entrepreneurship and to celebrate the more altruistic aspects of cartoons.

Of course, except for the occasional mention of Popeye, Thorp wrote solely about Disney characters. Thus, Warner Bros. made the attempt in the *Look* article to have its own product considered as part of the new mythology. Apparently included among those "less skillful cartoonists" discussed by Thorp who had not yet established their characters as products outside of their cartoons, the Warner Bros. animation studio may well have been attempting to show the marketing possibilities for its products by stressing their high moral content. Indeed, it was during the late 1930s that Warner Bros. moved heavily and successfully into cartoon star creation: Porky Pig first appeared in November 1935, Elmer Fudd and Daffy Duck in 1937, and Bugs Bunny in 1938. The article, then, refers to a constellation of marketing issues related to animation; just as with live-action filmmaking, the star system had become established not only as a means of regulating film budgets, but of regulating, in addition, audience interest in movies and in products from movies. Thorp correctly emphasized the public's desire to see star vehicles, and then to buy star products. As a result, Warner Bros., without the star stable of Disney or even Paramount (which produced "Popeye" and "Betty Boop" cartoons), asserted the worthiness of its film product as part of its attempt to create stars that could compete with Mickey Mouse and Donald Duck in all manner of merchandising.

Thorp's emphasis on the products of a specific studio points out the structuring absence in the *Look* article: Walt Disney. In 1939, one year after the triumph of the feature-length *Snow White,* no popular sources, from sociological ones like Thorp's to articles in magazines, questioned the quality, either aesthetic or moral, of Disney animation. Cruelty to animals, for instance, or excessive spitting would be unthinkable in a Disney cartoon. In part because of its unquestionably worthy content (and also because of its success at cornering the comic strip and nursery rhyme market) the Disney studio had established itself as the children's storyteller par excellence, as a worthwhile recipient of that commendation from Teachers College.

Warner Bros., clearly, wanted a share of the children's audience, perhaps not primarily to increase movie revenue so much as to build up profits from the marketing of cartoon characters. Because of this, the examples in the article stressed the wholesomeness of Warner's cartoons: no cruelty, no monsters, no taking of the Lord's name in vain, no sex. But Schlesinger's comment in the article, which could be read as an assertion of animation's broad appeal, also implied a certain indecision on the part of the non-Disney animation studios: "We cannot forget that while the cartoon today is excellent entertainment for young and old, it is primarily the favorite motion picture fare of children. Hence, we always must keep their best interests at heart by making our product proper for their impressionable minds."

Product Differentiation and Disney Domination

However inadvertently, Schlesinger must have been voicing the anxiety of most cartoon producers. How could they assert their similarity to the Disney product, which had come to define animation as child's fare, but at the same time differentiate their films from it? I do not mean to argue here that Disney appealed only to children, or that animation itself had come to be considered simply as children's entertainment; that would not happen until the full development of commercial television in the 1950s. I would argue, however, that Disney, at this time, achieved success by marketing his product as ideal children's fare, which prompted other cartoon producers to claim the same audience, but also to seek others, and in a manner absolutely different from how Disney himself found a niche with an older audience. Hence, while stressing children, Schlesinger also emphasized that they did not constitute the only audience for cartoons.

At the same time that non-Disney animation attempted to assert its fair share of the audience, Disney consolidated his hold over the child audience, and, by extension, certain sectors of the lower- and middle-class audience. Virtually simultaneous with the rise of Mickey Mouse as a movie star was his appearance in 1930 as a comic strip character. In hundreds of newspapers throughout the period I examine in this book, Mickey appeared in strips that were either based on his cartoons or original stories. Thus, Disney used comic strips to create interest in specific cartoons, and also to make Mickey (and, later, other characters created by the studio) either a daily or weekly part of people's lives.

Although it is difficult to determine precisely who read the comics during this period, we can perhaps guess at a wide child audience, and perhaps at a lower-class and less educated adult one. Certainly, popular mythology would imply that children thrived on newspaper funnies. One of the most enduring and endearing stories about Fiorello LaGuardia, for instance, is his taking to the radio waves during a New York newspaper strike to keep children informed on the adventures of Dick Tracy and other comic strip stars, with this arguably antilabor action by a progressive mayor nevertheless asserting his position as a government official who cared deeply about the daily lives of his youngest constituents. Furthermore, the easy transition of so many stars from cartoons to comics or vice versa (Popeye, Little Lulu, and L'il Abner are examples of the latter) at a time when cartoons were positioning themselves more and more forcefully as children's entertainment, also implies a largely adolescent audience for the funnies.

It is also possible that a lower-class adult audience, with less at stake in maintaining bourgeois literary standards, and which, because of its number of immigrants, was probably less proficient in English than the middle class, also would have enjoyed the comics. Indeed, that bastion of elite journalism, the *New York Times,* has steadfastly refused to run comic strips. As further circumstantial evidence, it is one of the jokes in *The More the Merrier,* a feature film from 1942, that two middle-class members of the federal bureaucracy—played by Charles Coburn and Joel McCrea—would take such delight in the Sunday comics, a delight that placed them squarely on the side of the common people.

Apparently in an attempt to attract that middle-class audience, in the early 1930s Disney began publishing "Mickey Mouse Movie Stories," a series of hardcover books containing stories from the cartoons and accompanied by drawings from those films. Throughout the decade Disney also placed a monthly nursery rhyme, again based on one of his films, in *Good Housekeeping* magazine, together with sketches of his characters.

In each instance, while certainly not excluding a lower-class audience, Disney was working within a network of bourgeois concerns and consumer habits. That the "Movie Stories" came out in hardcover rather than as comic books asserted their literary value, and *Good Housekeeping* served as the premier magazine of the middle-class homemaker who sought elegant recipes, information about domestic technology, and escape through proto-feminist fantasies about women traveling in Africa and Asia.

By catering to this more affluent audience, Disney claimed a very special status for his cartoons. Much in the manner endorsed by Thorp, they created a modern story-telling tradition. The *Good Housekeeping* nursery rhymes clearly were meant to be read by a parent, probably a mother, out loud to a child. The "Movie Stories" also seem designed for this purpose; while the language is not complex, it is also not simple enough for very early readers, and the beauty of the books themselves, testified to by Maurice Sendak in his introduction to a modern edition, seems certain to please the adult who would probably purchase the book and then use it to initiate the child into the wonders of literary experience.[9] Thus, Disney would accomplish what Schlesinger had asserted, in the *Look* article, to be the mission of all animation—the education and entertainment of America's youth.

Class, History, and Heterosexuality

While the apparent prohibition on kissing stated in the *Look* article seems extreme—characters kiss, however, chastely, in the cartoons from most studios throughout the classical period—there are significant differences from cartoon studio to cartoon studio in the representation of heterosexual desire. These differences hint at Disney's complicity with the kind of regulation of social norms expressed by the censorship code, and also hint at the attempts of the other studios to assert their differences from the Disney product, and, more broadly, at the efforts of the popular culture industry to explain contemporary historical events to a mass audience.

In the 1950s, cultural critics celebrated the modern era as one in which the privileges of the middle class had been extended more widely than ever before. As Jackson Lears has explained it, this period held out the possibility of a society in which "factory workers ate porterhouse steaks and drove Buick Electras."[10] But the project of constructing a universal middle class had begun well before mid-century, and at least as early as the period in which the United States moved closer to entering the Second World War. In attempting to create a consensus about what the United States might have to fight to defend, the popular culture industry frequently targeted perhaps the largest audience for any form of popular culture—the middle class. In doing so, they created a middle-class ideal that embodied all of the traditional virtues and values that would come to characterize the war aims in the United States, and that also were accessible to those who were unable economically to assert their own middle-class status. Rather than indicating anything conspiratorial on the part of popular culture, this particular equation of American values and the middle class demonstrates one of the complex wartime interactions between industry, the government, and the audience that I will explore at greater length in Chapter 5.

Disney cartoons from the 1930s and '40s typically represent one of those values—heterosexuality—not so much to animate desire as to assert a status quo ideology about class, either to smooth over Depression-era tension or to create a consensus about the United States' entrance into World War II. At about the time of the *Look* article, with the United States moving toward an ideological position justifying participation in the war, any number of Hollywood feature films began to stress an "ideology of affirmation," one in which "the goal of commitment has to be so evident, so natural, so necessary, that it needs no ulterior justification."[11] In examining class, then, the 1940s feature film narrative created a culture in which we could be superficially different because we were fundamentally the same. The cartoons I viewed from the war period, and particularly the Disney cartoons, created the same homogeneity, but did so much less subtly; the lower class tended to disappear, and the middle class seemed to be everywhere, defining itself not so much by economic interests as by romantic ones.

Various Disney cartoons made throughout the classical period exemplify the confluence of class issues and what Hollywood considered normal sexuality. *The Cookie Carnival,* a "Silly Symphony" from 1933, shows the possibilities for a typical movie hobo from the period to climb the social ladder and also to marry a beauty queen. Two decades later, Disney's history of music, *Toot, Whistle, Plunk, and Boom* (1953), "proves" that Western notions of high culture stem from heterosexual desire, as men throughout history are shown inventing music and musical instruments in order to attract women. Then, in another representation of the creation of Western mainstream culture, *Noah's Ark* (1959), the Bible story is set against the melodrama of the hippopotamus pair, who work out their marital problems during the flood.

Mickey Mouse, however, from the late 1930s on, functioned as the principal animated incarnation of this concern with sex and class. *Hawaiian Holiday* (1937), for instance, shows the possibility for travel to exotic locales of absolute leisure for the romantic couple, in this case Mickey and Minnie. In *The Brave Little Tailor* (1938), Mickey as the title character in the fairy tale captures the giant in order to assure the safety of the king, and to guarantee his own class ascendancy through marriage: he wins the hand of Princess Minnie.

Yet another "Mickey Mouse" film fully demonstrates the attempts during the war period to create a national consensus around these issues. *The Nifty Nineties* (1941) places Mickey and Minnie at the turn of the century, an era that became increasingly important to the American cinema of the period (it served as the setting for such feature films as *Strawberry Blonde, The Magnificent Ambersons, Gentleman Jim,* and *Meet Me in St. Louis,* and also for significant portions of *Citizen Kane*).

The credits are typical of Mickey's films from the period. A giant closeup of the star, the star's name, Disney's name, and a brand name—Technicolor—dominate the screen. Rare for Mickey's films from the time, however, there is also a song in the background, one that immediately informs us, even before the title, of the film's historical setting: "A day in the nineties when Grandma was a girl / The horseless carriage was quite the show / Grandpa fussed when the thing wouldn't go . . . / Come take a look / In our picture book." Before the narrative proper has begun, the song has performed several functions. It celebrates traditional courtship, in which a man operates a car while making overtures to a woman. Moreover, it has implicitly placed that behavior within a social and economic class—Grandpa owns an automobile. Finally, it has invited the audience to engage in familial nostalgia, as the viewers are asked to look at a scrapbook. The film thus displaces the values that Mickey and Minnie embody in the film to all of the people—the audience that makes up the vast family for *The Nifty Nineties*—watching the film.

Following the lyrics, the credits show a black-and-white photograph of Mickey, as the grandpa of that family, standing next to the horseless carriage. The photo then changes to Technicolor and comes to life. Audiences in 1941 may have been witnessing the beginning of a new element of film language or, at least, were being presented with a new meaning for that element. After *The Nifty Nineties,* at least two films—*Meet Me in St. Louis* and *Gentleman Jim,* both of which take place in the same era—would employ some variation on this scrapbook-come-to-life motif (the former, for instance, makes use of a sampler). In each case, this serves to historicize the narrative to come. That is, the film performs an archeological function—uncovering that which had been forgotten—rather than a more properly fictional one of making up a story. This increased reality effect then works to validate the values of the films, which in each case either celebrate a middle or upper middle class (*Meet Me in St. Louis*) or celebrate the opportunities for members of the lower class to experience rapid upward social mobility (*Gentleman Jim*). As with *The Nifty Nineties,* the scrapbook places all of this within a family context; the audience members are not so much watching some stranger's story as watching a story about their own ancestors.

Once the viewers have entered this historical/familial space and thereby been transformed from a diverse audience into a homogeneous one, they hear a turn-of-the-century celebration of leisure and romance—"Strolling in the Park One Day"—sung in the background. Mickey and Minnie, both nattily dressed, enact the song's courtship ritual. Later, as they go driving in the park, they pass Goofy on a bike, and then Donald and Daisy with their nephews. The film has turned familiar Disney characters who, in 1941, had been created just a few years before, into American historical traditions; they are instantly recognizable icons that go back at least half a century.[12] They also enjoy absolute leisure, and in the case of Minnie and Mickey and Donald and Daisy they have formed heterosexual couples and have begun families (the duck-nephews). Finally, through their dress and their free time, they clearly enjoy a certain class status. *The Nifty Nineties* told viewers exactly what they soon would be fighting for: a shared culture based on families, leisure, and heterosexual couples, represented by beloved Disney characters whom the film turns into enduring middle-class icons.

Though the film stresses typical Hollywood heterosexuality, the legal documentation for the cartoon insists upon female sexual desire as the impetus for the narrative. As a result, the copyright information indicates the possibility of a 1940s audience viewing the film not simply as an homage to patriotism and tradition, but as the fulfillment of a heterosexual male fantasy about sexually aggressive women. At the begining of the cartoon, as Mickey strolls through the park, he passes by a sign for the "Bloomer Girls" baseball team. The poster shows a lineup of conventionally attractive women with 1890s hourglass figures, and Mickey tips his hat to them. Then he meets Minnie, who, as part of the period courtship ritual, purposely drops her handkerchief. With the effective shorthand of the seven-minute narrative form, these opening shots of the cartoon establish that both Mickey and Minnie are on the make, looking to meet someone.

The copyright information in the Library of Congress, however, describes Mickey as simply "skipping through a park" when "Minnie appears." She is "walking coquettishly, and drops her handkerchief" (fig. 2). Making no mention of Mickey's interest in the Bloomer Girls, the description implies that the male mouse is the dupe (however willing) of Minnie's flirtatiousness. The film itself depicts the main characters as equally interested in and responsible for their courtship. But the legal representation of the film—the copyright information—constructs a plot upon the narrative impulse the Production Code quite specifically attempted to limit, if not prohibit: female sexual desire. Though the language today might seem innocent enough, the copyright description uses a term—coquette—a stronger variant of which—cocotte—was in a group outlawed by the Code along with "chippie," "broad," and "alley cat."

As a result, a conventionally wholesome cartoon about the links between heterosexual romance and class privilege becomes in this legal narrative a story that probably could not have passed muster at the Hays Office. Indeed, the copyright description provides no indication of the characters' class status—which the film asserts instantly, with the introductory song—until fairly late in the synopsis, when Mickey's car appears. Legally, then, the film concerns sexuality far more thoroughly than does the narrative constructed by the images and the soundtrack, in which heterosexual romance functions to maintain and historicize a class-based status quo.

From this, we can see the contradictions in the various legal discourses that governed film practice. The standards adhered to by the film studios—the Production Code—forbade overt displays of sexuality. The legal document registering *The Nifty Nineties,* however, foregrounds it in reconstructing the film's narrative. Michel Foucault has written that "If sex is repressed, that is, condemned to prohibition . . . then the mere fact that one is speaking about it has the appearance of a deliberate transgression," and this is precisely the operation here. Rather than simply condemning sexuality to silence, the law itself becomes transgressive and extralegal.[13] The apparatus that represses also offers the guilty pleasure of subversion.

Additionally, the legal narrative of *The Nifty Nineties* comes into

Figure 2. Minnie Mouse-as-coquette in *The Nifty Nineties* (1941).

direct conflict with the film itself. A Disney cartoon that conforms completely to censorship codes and dominant 1940s ideology comes to be described legally in a manner that violated the letter of the former, and perhaps the spirit of the latter. This may tell us as much about reception as it does copyright practice, and also something about the popularity and the genius of the Disney product. Despite the apparent conservatism of the films, they could still easily be read, as in the case of the person (probably a Disney employee) who wrote the copyright description, in a manner that foregrounded that which apparently was safely submerged in the film text: sexuality, and specifically female sexuality.

Race, Gender, and the Cartoon Harem

Mickey Mouse, of course, was nothing if not wholesome, and so it seems appropriate that sexuality in his films had to be read between the lines rather than conclusively demonstrated. In other Disney films in which Mickey does not appear, however, sexuality seems to be precisely the issue, despite censorship restrictions. In part, this mode of representation could be made acceptable because of the manner in which sexual stereotypes reinforced racist ones and could be mobilized to parody and also normalize a typical male response to female sexuality. The Near and Middle East, the region, following Edward Said, that we can refer to as the Orient, and even more specifically within this region, the harem, functioned as the signifier of hyper-heterosexuality that masked both racism and ethnocentrism.

Along with an animated Africa, the Hollywood version of the Orient worked as one of the great sites of otherness not only in cartoons but also in live-action feature films. During eras of colonialism as well as decolonization, a variety of cartoons insisted on the irrationality and exoticism of the region: *Aladdin and His Wonderful Lamp* (1934), *Sinbad the Sailor* (1935), and *Ali Baba* (1936), all produced by Ub Iwerks for Celebrity Pictures; or the Warner Bros./U.S. government training cartoon co-production, *Payday* (1944); or *Popeye the Sailor Meets Sindbad the Sailor* (Paramount, 1936), or Superman in *The Mummy Strikes* (Paramount, 1943), Bugs Bunny in *Sahara Hare* (Warner Bros., 1955), and Woody Woodpecker in *A Lad in Bagdad* (Universal, 1962).

Within these cartoons, the use of the harem-as-signifier of a vast geographical, cultural space functioned much as it had since well before the nineteenth-century colonial adventures. Through an influx of postcards and book illustrations, by the turn of the century most Westerners were familiar with the primary tropes of the harem: the concubines, the opium pipes, and the general decadence. In an examination of these images, Malek Alloula has explained that most harem photographs were taken by European photographers, and that the poses themselves were modeled after paintings by Jean Ingres or Eugène Delacroix. Alloula argues that the harem postcard constitutes "a naive 'art' that rests, and operates, upon a false equivalency (namely, that illusion equals reality). . . . It literally takes its desires for realities."[14]

During the first years of the cinema, Thomas Edison capitalized on this late nineteenty-century fascination with these images by filming *Ella Lola's Turkish Dance*. Several years later, around the beginning of the movies' classical era, D. W. Griffith, perhaps the leading Victorian sensibility of the silent period, articulated the Edwardian obsession with exotic sexual spheres through the harem images in such films as *Intolerance* (1916).

By the 1930s, this representation of the harem had become standard not only in adult entertainments such as Griffith's, but also in those designed for a general, or even an adolescent audience. Although Walt Disney had established his hegemony over the child audience, this control did not preclude his product from exemplifying the orientalist approach to the harem, as we can see from the "Donald Duck" cartoon, *A Good Time for a Dime* (1941). Because it is a cartoon, *A Good Time for a Dime* cannot meet the postcards' standard of signalling reality. But the cartoon demonstrates nevertheless the possibilities for representing objectified female sexuality even to children as long as that representation conforms to accepted standards of racism and misogyny.

In the cartoon, an amusement park advertises "A Good Time for a Dime," and Donald, ever the gullible consumer, decides to try his luck. He walks into a room lined with Edison-style kinetoscopes, each one designed for an individual viewer. Most of them have travelogue labels, so Donald puts his money into a machine listing the "Dance of the Seven Veils" as its subject. For a few moments Donald's point of view becomes the spectator's as he and we watch Daisy Duck, dressed in Hollywood's version of harem clothes, perform the dance. The film accepts this dominant male spectatorial position, first analyzed in detail by Laura Mulvey, and then parodies it; the machine breaks down and Donald becomes furious (fig. 3).[15]

For those interested in the cinema's identificatory practices, this alignment of spectator and character and then its break up—we laugh at Donald's fury—may well stand out as the furthest the classical cinema will go to destabilize this type of male, voyeuristic vision. In a different context, however, the entire scene unproblematically represents all of the Western stereotypes about the harem. *A Good Time for a Dime* manufactures the harem as a setting separate from any social phenomena and full of obliging women happily catering to male sexual fantasies.

A Good Time for a Dime reproduces the trope of the secluded harem woman in the most concentrated form possible. Hanna Papanek writes that because secluded women like those in the harem do not have the opportunity to develop public systems of expression, the West cannot easily obtain access to their points of view, and has come to depend simply on "the writings of men through the ages."[16] As an example of that denial of a complex harem viewpoint and of the cinema's adoption of standard representational practices in regard to the harem, the Disney film provides only one point of view—Daisy occasionally looks directly back at the spectator, in the manner of the women in the postcards.

Figure 3. The frustrated male spectator in *A Good Time for a Dime* (1941).

Rather than disrupting a voyeuristic fantasy by calling attention to it, Daisy's acknowledgment of the viewer, like that of the postcard women and Edison's Ella Lola, institutionalizes this viewing process as nothing to be ashamed of, as a standard relationship between male and female, between West and East; that is, the harem woman must be aware of her position as both sexual object of desire and colonial object of domination.

Like most Hollywood cartoons from the period, *A Good Time for a Dime* lasts a little longer than six minutes. The "Dance of the Seven Veils" sequence, from the moment Donald sees the kinetoscope to the end of Daisy's performance, runs about forty-five seconds. Visually, with its film-within-a-film structure, the performance is the most interesting segment of the cartoon, and its placement at the beginning underscores its importance as a means of piquing spectator interest. But the second sequence in the film (Donald's attempt to win a prize from an "Iron Claw" machine) lasts over two minutes, and the third and last sequence (Donald's disastrous "Thrill of Aviation" joyride) runs almost three minutes.

In terms of the film's legal status, however, as an artifact eligible for copyright, the "Seven Veils" sequence is of virtually equal importance to the sequences coming after; the viewer who described the film devoted slightly less than twelve lines to Daisy's dance, but only fifteen to the lengthy second sequence, and somewhat less than that to a ninety-second segment of the third. I do not mean to make an argument here that equates narrative and ideological significance with space on the copyright form. In a manner similar to *The Nifty Nineties,* however, where the film viewer/describer concentrated on Minnie's coquettishness, here the sexualized behavior clearly captured a disproportionate amount of the viewer's attention.

Like the copyright description of *The Nifty Nineties,* the one for *A Good Time for a Dime* provides us with at least some information about a cartoon's possible reception, as it codifies a type of male spectatorship. In so doing, the description shows the confluence of, and cooperation between, the law (of copyright), technology (not only the cartoon, but the kinetoscope within the cartoon), and a specific subject/object relationship (men watching women). In those sections of the description dealing with the "Seven Veils" sequence, the viewer/describer identifies Donald's moods and behavior in far greater detail than in any other segment of the copyright. The beginning of the description acknowledges the magic of the machine in capturing the male spectator and riveting his attention: "He starts to pass THE DANCE OF THE SEVEN VEILS, but suddenly realizes what the title implies." Then, he "inserts his coin in the machine and the light goes on," with the use of the verb "insert" and its various connotations establishing the tone of the rest of the description. Donald "is greatly excited by Daisy's dance and whistles as she starts to discard her veils." He then "settles into a pose of extreme interest." When the image accidentally turns upside-down, "Don climbs on top of the machine," itself an image not free of sexual currency, and then, as Daisy drops "the third and fourth veils," he "settles into a relaxed position." When the machine breaks down, Donald becomes "exasperated," and then shakes it "violently."

The very subject of this sequence in the cartoon—a woman removing her clothes and a man watching her—provoked the viewer/describer to write almost exclusively about male response, thus making that response the legal reading of that segment of the film. In the second section of the description, about the iron claw machine, Donald is described as "satisfied," and in the third, about the airplane ride, he is "thoroughly enjoying himself," before being thrown into a "rage." Throughout the film, Donald shows the usual range of his emotions; a viewer would be hard-pressed to prove that he exhibits more of his behavioral tics in the "Seven Veils" scene than in any other. But because that scene deals so specifically with a male voyeurism so important to the Hollywood cinema, and which, frequently, is designed to elicit versions of the behavior that Donald exhibits, that behavior comes to be not necessarily the correct reading of the film, but certainly the legal one. In addition, the description establishes the harem, or, more properly, the representation of the harem, as that which naturally provokes that kind of behavior; a "greatly excited" male spectator becomes an acceptable cartoon character because of that which the male spectator views.

Representational Practice and Legal Repression

Despite such occasional forays into the exotic as *A Good Time for a Dime*, for the most part during this period Disney concentrated on representing American culture in the manner of *The Nifty Nineties*. Rather than changing his subject matter, Disney instead, for a brief period in the late 1940s, adapted some of the representational strategies of competing studios to attract a newly affluent teenage audience that had come to make up the largest portion of the filmgoing public.[17] Disney's jazzed up cartoons, while frequently adhering to his previous films' interpretation of American culture, also made overt references to sexual behavior, and not in ducks or mice but in seemingly realistic women.

During the war period, Columbia and MGM introduced pin-up girls to some of their cartoons, the former with Daisy Mae in the short-lived "Li'l Abner" series, the latter with Red Hot, the chanteuse who appeared in a number of Tex Avery's animated shorts (fig. 4). Avery has admitted that these cartoons had trouble with Hollywood censors, who were concerned not so much with the representation of Red as with the depiction of the male response to her. In typical Avery fashion, the wolf who watches Red in these cartoons has his eyes bug out, steam come from his collar, and his ears stand erect. According to Avery, when the representative from the Hays Office watched *Red Hot Riding Hood* (1943), he said, "Boy, he's getting too worked up," and required a few cuts in the film. A member of the military brass asked for an uncut version to show to his men, and, again according to Avery, "it went great overseas."[18]

Figure 4. Red Hot and the Wolf in *Swing Shift Cinderella* (MGM, 1945).

I will examine the issue at greater length below, in a discussion of wartime "Superman" cartoons, but, apparently, the Production Code during the 1940s was flexible enough to tolerate what might be considered sexier films if, however tangentially (as in the case of *Red Hot Riding Hood*), they could be deemed to be beneficial to the war effort. Taking advantage of these new standards, in the feature-length compilation film *Make Mine Music* Disney included, along with some rather traditional and sentimental segments, a sequence called "All the Cats Join In," in which teenagers dance to Benny Goodman's music at a malt shop.

At the beginning of the sequence we see perhaps the most explicit representation of the female body in the Disney canon. A teenager, getting ready for a date, slips out of her clothes and steps into the shower, and we can clearly see the outline of her body through the shower door. As I discuss in Chapter 4, critics objected to the implication of teenage sexuality in the film, concentrating on the gyrations of the jitterbug at the malt shop rather than on this shower sequence. Partially as a result of these objections, within a few years Disney had found a way of depicting women who seemed realistic, but in much less charged settings: consider *Cinderella* (1950), for instance, or *Alice in Wonderland* (1951).

Copyright documents for *Make Mine Music* indicate, moreover, that the Disney studio may have felt some dis-ease about "All the Cats Join In," and may have tried to contain any hint of sexuality through the language used to make the film legal. With *The Nifty Nineties* and *A Good Time for a Dime,* copyright material revealed the possibility of reading the films at least partially against the grain, emphasizing Minnie's sexual behavior, for instance, or devoting a disproportionate amount of space to Daisy's Dance of the Seven Veils. *Make Mine Music* demonstrates the same tension between the film and its legal incarnation, but in a manner that indicates the power of the latter to control the former.

Briefly and to the point, the description of "All the Cats Join In" stresses the sequence's graphic design, and turns the story into something of a teenage idyll:

> An animated pencil begins to draw in a sketch book. Draws juke box, then cat. Erases cat, draws boy. Boy calls girl over phone, she hears music from juke box, makes date. Proceeds to get ready, is imitated and heckled by little sister. Pencil draws jalopy and boy; girl joins him and they pick up kids in neighborhood. All go to malt shop; order fancy sundaes, play juke box and dance.

The description indicates something of the complexity of the period's filmmaking practice. Columbia and MGM, in order to compete with Disney and to attract an audience that Disney perhaps did not yet control, adopted a strategy that intensified the representation of female sexuality and male sexual response. Disney, himself competing with other studios and also in search of new audiences, borrowed at least tentatively from MGM/Columbia. Showing the conflicts within Disney practice itself, however, the legal representation of the film stresses the familial and the old-fashioned.

As the example from *Make Mine Music* demonstrates, sexuality never consists of a coherent system that includes all representational systems and all manner of social discourse. Rather, as Richard Dyer has pointed out, it consists of "clusters of ideas" which frequently come into conflict with each other.[19] Thus, rather than the depiction of the teenager in "Make Mine Music" indicating the period's apparently more open views toward sexuality, views that would lead to the popularity of Jane Russell and Marilyn Monroe and to the best-selling status of Alfred Kinsey's *Sexual Behavior in the Human Male,* the film and the legal document demonstrate the manner in which sexuality could be constructed and repressed at the same time, and in relation to the same artifact. In the marketplace, *Make Mine Music* made ample and obvious uses of female sexuality; for legal purposes and also for its place in the archive—the Library of Congress—the film was made absolutely safe.

Betty Boop's Body

Other cartoon producers faced the same Production Code prohibitions on sexuality in general and female sexuality in particular. But they also faced the manner in which Disney dominated animation. As a result, Max and Dave Fleischer, who released their cartoons through Paramount, opted for a series of representational extremes during the 1930s and early 1940s, and the shift from one to the other hints at the possibility to defy the Code as well as to the potential sternness of the censorship system. These shifts also coincide with some of the scientific findings about the influence of motion pictures. Moreover, they hint at changes in film production during the period, changes that encouraged a politically aware cinema that, by the 1940s, extended even to animation, and that worked to replace sexual concerns with more purely social ones.

In the early 1930s, the Fleischers and Paramount sought to construct a new star, Betty Boop, through female heterosexual desire *tout court,* without any connection to class issues, as in Disney films, and with barely any relation to the cartoon narrative. Early in her career, Betty functioned as much as a brand name as a performer. In *Betty Boop's Ker-Choo* (1933), for instance, the star is inscribed and reinscribed within the credits. Superimposed on the Paramount trademark—the mountain and clouds—a proscenium arch is labeled: "Betty Boop Cartoon." Even more than being a Paramount product, the film seems to be a product of the star herself, a creative power typically denied to live-action performers. Beneath the proscenium, we read "Max Fleischer Presents Betty Boop's Ker-Choo," after which we see a closeup of the star. She says "Yoo- hoo" and winks right at the spectator while a male voice sings, "She can win you with a wink," and then, "Ain't she cute?" Betty's name is written under the closeup (fig. 5).

Numerous "Betty Boop" films from this period begin in the same manner, or in a variation thereof. These cartoons celebrate Betty herself as

Figure 5. From the credit sequence of *Betty Boop's Ker-Choo* (Paramount, 1933).

synonomous with film entertainment—with her name on the proscenium—and, indeed, as the virtual creator of it. Then, she is aligned with the individual film, as her name frequently appears in the title. Just as she becomes the brand name that encompasses all of her films, however, so too does she take on a different kind of extradiegetic status. Betty seems to exist outside of the narrative proper as a star appearing in the film, as someone who acknowledges the spectator. More exactly, she acknowledges the male spectator, whose responses to the star are directed by the male voice, and whom Betty beckons with her wink and "yoo-hoo." Her films, then, even in the credit sequences, create a relay of male spectatorial desire in which Betty's body serves as the trademark of the series just as does the Paramount mountain, and also serves, through Betty's come-hither look, as an invitation to the postcredit film narrative.

Issues closely related to Betty's gaze had entered the scientific discourse about the movies' effects at about the same time as the cartoon star's debut. In 1933, for example, in a *Movies and Conduct* chapter called "Imitation by Adolescents," Herbert Blumer cited the testimonial evidence of two college women:

> Although I have never adopted any ways of flirting, I do not mean that I have never tried out the technique of the stars. Once I decided to try out a type of eye work that I had admired in a movie. It worked so well that I have not dared to use it since.
>
> Now here's a real confession! Ever since I saw Joan Crawford use her eyes to flirt with people, I caught that trick and use it to good advantage.[20]

In the role of objective scientific observer, Blumer refused to state an opinion on adolescent imitation of all manner of movie behavior, although he did assert that "much is taken over and is incorporated into conduct."[21] Yet the evidence itself indicates the equation between female heterosexual desire, the female gaze directed at men (Blumer stressed throughout that the learned behavior was always used in relation to the opposite sex), and danger. One woman, apparently astonished at the power invested in her look, did not "dare" flirt in such a manner again, while another related her own experience in the context of a "confession."

The Catholic Legion of Decency and other procensorship groups undoubtedly adopted this same apocalyptic discourse in talking about the effects of the cinema, and it is largely this representation of female sexuality that the Production Code sought to curb once the studios agreed to adhere strictly to it. Indeed, changes in the credit sequences of "Betty Boop" films indicate that the Fleischer brothers responded to the requirements of the Production Code while still working to construct female sexuality as a selling point.

Betty Boop and the Little King (1936) exemplifies this practice. As a tribute to her star power, Betty was often used to sell other cartoon stars by appearing in movies with them: besides this one with the Little King, a character that originated in *The New Yorker* and then moved to newspaper comics, the Fleischers produced *Betty Boop with Henry, the Funniest Living American* (1935) and *Betty Boop and Little Jimmy* (1936). And in these cartoons from the middle period of her career, Betty once again becomes a trademark and brand name for her films. The credits show her silhouetted against a window, creating something of a proscenium effect, with "Betty Boop Cartoon" written at the top, and then "Adolph Zukor Presents a Max Fleischer Cartoon." Thus, Betty's body has taken the place of the Paramount mountain of earlier cartoons as the signifier of the kind of film to come. But as if to defuse the character's sexuality even while stressing it, her face is obscured; because of the silhouette, it is difficult to tell whether Betty faces the camera or has her back to it (fig. 6). This denial of her gaze at the spectator denies the relay that Blumer examines, in which the female star's look at the male character influences the female spectator to "dare" to attempt the same approach.[22]

Perhaps to placate those censors who, according to *Look* magazine, preferred "romance of the hand-holding type," these same "Betty Boop" films often constructed a postcredit socio-sexual sphere that functioned much like that of the Disney cartoons. Celebrating class distinctions, these films emphasized the problems of the wealthy, and implied that they could be solved by contact with a less restrained working class. In *Betty Boop and the Little King,* for instance, Betty performs in a vaudeville theater until she is discovered by the king, who has left an opera house to go slumming. The film has turned Betty into an object for an appreciative male gaze—the king's. Then, in a happy ending, the Little King, having escaped from a stultifying upper-class setting and from his much taller queen, discovers with Betty the rewards of low-brow entertainment and lower-class sexuality.

Particularly after the major studios' commitment to the Production Code in 1934, numerous "Betty Boop" cartoons resist narrativizing the sexuality of the credits, and instead concentrate on using the star to assuage class tension. In *Little Nobody,* for instance, an apparently middle-class Betty (she lives in a modest house) cements a friendship with the formerly nasty wealthy woman across the street when Betty's dog Pudgy rescues and then romances the neighbor's pet. *Musical Mountaineers* (1939) shows the humanizing effects that a citified Betty has on hostile, ill-mannered, poorly-spoken hillbillies.

Discussing the relationship between censorship and the British silent cinema, Annette Kuhn has pointed out how "discourses around the body and its sexuality . . . are instrumental . . . in producing . . . certain forms of knowledge: namely, knowledges which aspire to order the domain of the sexual as it participates in and is contained by the social, and which constitute the body and its sexuality as essentially social processes."[23] During the sound period and in a different country, but confronted by similar censorship pressures, "Betty Boop" cartoons performed the same function. That is, after asserting product differentiation—through brand name and the representation of female sexuality, the credits told the

Figure 6. From the credit sequence of *Betty Boop and Little Jimmy*, produced after stricter enforcement of the Production Code began (Paramount, 1936).

viewer over and over again that these were not Walt Disney films—they then had to control Betty's body. If the credits allow Betty a potentially subversive power denied practically all other female performers from the period, the narratives restrain that power by insisting on the link between that body and the maintenance of a well-defined, class-based status quo.

Betty in the Archive

Documents registering "Betty Boop" films in the Library of Congress indicate, in a manner similar to the films themselves, some of the differences between products from Disney and Paramount. The Disney copyrights provide descriptions—that is, interpretations of actions and character motivation: Minnie might emerge as a coquette in the copyright, or Donald's moods would be analyzed in some detail. In contrast, Paramount simply copyrighted a cartoon's dialogue. In being preserved for posterity, for instance, *Betty Boop's Ker-Choo* becomes forty lines of dialogue without overt clues as to who speaks them. Rather than resembling a script, the copyright document looks like a poem, partly in iambic pentameter, partly in free verse. Only if one has seen any "Betty Boop" films is it possible to determine that the rhymed segments comprise the theme song at the start and then, later, Betty's own performance.

In the films made before the stiffening of the Production Code in 1934, this strategy for transforming cartoons into legal documents underscores the link between Betty-as-sex symbol and Betty-as-product. Once again from *Betty Boop's Ker-Choo,* the first eighteen lines of copyright record the opening song, sung over the credits and discussed above: ". . . Made of pen and ink / She can win you with a wink—yoo-hoo / Ain't she cute / Boop-oop-a-doop / Sweet Betty." The final twenty-two lines of the copyright, some of them as short as one word ("Gesundheit" or "Ker-Choo") provide the dialogue for the film that comes after the credits.

Like the description of *A Good Time for a Dime,* in which the viewer/author seems to devote much more attention to the "Seven Veils" number than would be required by the actual length of the scene, so too does this document for *Betty Boop's Ker-Choo* dwell inordinately—almost half of the manuscript—on a fairly short credit sequence. As a result, that which gets the most legal attention is that which the films themselves, in competition with Disney cartoons, sought to sell the hardest—Betty's sexuality; her direct appeal to the spectator, and more specifically to the male spectator, which comes before the beginning of the narrative proper.

By so concentrating on this credit sequence, the copyright documents indicate Paramount's own lack of interest in making the films themselves comprehensible. Indeed, without any cues as to who speaks, the dialogue often makes no sense at all: "One, two—where's your driver? How I should know? As usual. Why are you late? I got a 'code' in my 'doze.'" Rather, the documents demonstrate Betty's own importance, and the desire to protect her status as a product owned by Paramount. Just as the credit sequences in these films overdetermine Betty—by placing her name over the Paramount trademark, frequently using her name in the title, and presenting her, in closeup, to the audience—so too do the documents stress the importance not of the film itself, but of Betty as the star who appears in the film. In the copyright document, it is Betty Boop, and, more specifically, Betty's signification of female sexuality (apparent in the theme song full of double-entendres) that makes the film not only intelligible, but also something worth protecting legally. Betty's star image dominates the copyright document, and it is a star image constructed largely through sexual imagery, as the opening lines of the copyright inform us that "A hot cornet can go— / But a hot cornet can't boop-oop-a-doop / Like Betty Boop can do."

"Betty Boop" copyright material from the post-1934 period points out the different strategies of these films themselves, and also show, in a manner similar to the late-1940s Disney cartoons, how legal discourse helped to contain the representation of sexuality. Undoubtedly because of the more strictly enforced Production Code, these "Betty Boop" cartoons began only with music, and, as mentioned above, with Betty herself silhouetted against a window rather than winking at the spectator. Thus, the credit sequence, so important to the copyright documents just a few years before, disappears from the later papers. Just as with the copyright from *Betty Boop's Ker-Choo,* the document for *Betty Boop and the Little King* also starts with a song. But rather than asserting Betty's irresistibility, this one, recording the first postcredit sequence in the film, parodies the high culture of opera, as the document transcribes the diva's nonsense aria: "Ta A Noi, A Noi Sara, Ha Ha Ha Ha Ha Ha La Vi."

If the beginning of the document ably demonstrates the changes wrought on Betty by a stricter censorship system, the end shows how the Paramount cartoon copyright practice, which remained unaltered throughout the decade, could make the film, as legally documented, less potentially objectionable than the film as shown in theaters. As I have discussed, *Betty Boop and the Little King* ends with the clear implication that Betty has become the king's lower-class kept woman; in a sequence with no dialogue we see the king and the queen in their limousine, with Betty on the runnning board giggling as she holds the male monarch's hand. The copyright description, however, ends practically as it begins, with a transcription of the king's response to seeing Betty on stage and also of the crowd's reaction: "Ohhhh! My oh my. Ha ha ha ha."

This symmetrical document, which begins and ends with laughter, shows the inadvertent strategy of the Paramount system of copyright. By recording dialogue, the document, at the beginning, faithfully reflects the changes in the cartoons, changes designed to lessen the threat to accepted standards posed by Betty. At the end, however, that same recording of dialogue provides for an entirely different interpretation of the film than does the film itself. That which even the Production Code permitted—Betty's kept-woman status—the legal document, by concentrating on dialogue and neglecting visual representation, forbade, turning any hint of prurience into absolute innocuousness.

The Shifting Body

Betty's appearance changed considerably over the 1930s, at least in part in response to the Production Code.[24] The Fleischers still emphasized her curves, visible under her clothes, and still looked for any chance to show her legs. But as Leonard Maltin has noted, Betty's skirts got longer and her décolletage less extreme.[25] Despite the changes, in the same year as the *Look* article, 1939, and demonstrating the tension between even Betty's relatively sedate appearance and the industrial and social pressure to control the representation of female sexuality, Betty stopped appearing in cartoons. In her place, the Fleischer studio attempted to create a new star in Gabby, the Liliputian who had appeared in the feature-length cartoon, *Gulliver's Travels,* in 1939, and who began appearing in cartoon shorts in 1940 (fig. 7).

Gabby's diminutive size and squeaky voice posit him as something of an adult-adolescent, and his films concern not so much heterosexual relationships as hierarchical ones taking place in an all-male universe; Gabby and the king in *King for a Day* (1940), Gabby and the mayor in *It's a Hap-Hap-Happy Day* (1941), or Gabby and the young boy in *Gabby Goes Fishing* (1941).

The shift in production at the Fleischer Studio, from Betty Boop to Gabby, demonstrates the epistemological shift throughout the 1930s and early 1940s in discourses about the body. In the mid to late 1930s, the more stringent enforcement of the Production Code helped lead to the popularity of Shirley Temple, with the female child's body emerging as a safe place for working out Depression-era tensions about class, upward mobility, and family, in addition to sexuality.[26] Over a longer period of time, the pre-Production Code Betty Boop yields to the one from the post-Production Code period, and then, finally, gives way to a male star, and something of a childlike one at that. With Betty, the field of the sociosexual gradually comes to reinforce the class hierarchy, with the Fleischer brothers' product finally replicating Disney ideology despite the conscious opposition to the typical Disney cartoon. With Gabby, even romance of the "hand-holding type" is eliminated, and the emphasis

Figure 7. Gabby, who appeared in cartoons produced by the Fleischer Brothers for Paramount Pictures from 1939 to 1941.

rests solely on the assertion of the rights of elected and monarchical officials like the mayor and the king to rule, and the necessity for adult-children like Gabby to submit to their authority.

This does not indicate a linear evolution toward absolutely safe sexual relationships, or single-sex relationships like those in the "Gabby" films that typically have nothing to do with sexual desire. After the major studios' commitment to the Production Code, Ub Iwerks, an independent animator who had worked with Disney, made cartoons that at least implicitly sexualized male relationsips, even if only to show their apparent abnormality. There is, for example, a stereotypically gay pirate who is attacked by his shipmates in *Sinbad the Sailor,* from 1935. In addition, even during the latter stages of Betty Boop's transformation and just a year before the *Look* article that insisted on the "hand-holding" innocence of the Warner Bros. cartoon product, that studio produced *Jungle Jitters* (1938), in which an ugly white queen ruling over a tribe of black natives demonstrates the horror of female heterosexual desire by virtually imprisoning the first available nonblack suitor.

From the Sexual to the Social

Throughout the 1930s, then, although the cartoon studios publicized their emphasis on "hand-holding," they portrayed, at least in parody form, other kinds of relationships. But World War II wrought several changes on cartoon representations of the body and of sexuality. As discussed above, some studios, most notably MGM (Tex Avery's cartoons) and Columbia ("Li'l Abner") intensified female sexuality, and often justified the practice by associating it with the war effort. The Fleischer brothers, after stopping production on "Betty Boop" films in 1939, produced "Gabby" cartoons only in 1940 and 1941, and then, beginning in 1941, began the "Superman" cartoon series. Thus, the star's body, which had been the object of male desire (Betty), changed first into a perpetually childlike, desexualized male (Gabby), and then into a superman working to rid the United States of all enemies, foreign and domestic. The socio-sexual sphere of the early 1930s became the primarily social sphere of the war period, in which all desire must be sublimated to one's patriotic duty (fig. 8).

The Fleischer Studio and then the Famous Studio produced seventeen "Superman" cartoons for Paramount between 1941 and 1943. Many of them, predictably and understandably, posit two primary threats to the United States, one from the Japanese and another from the Germans. *The Japoteurs* (1943), for instance, confronts Superman with a Japanese émigré in the United States who seeks to hijack a new bomber to Tokyo, and in *Jungle Drums* (1943) the scene shifts from the United States to Africa for a struggle between super races, Superman's and Adolf Hitler's. In the cartoon, Germans have gained control of a black African tribe, using the natives to create a front for Nazi activities on the continent. Reporting for *The Daily Planet,* Clark Kent and Lois Lane discover the operation. Lois is taken captive, after which Superman rescues her and

Figure 8. Between 1941 and 1943, the Fleischer Brothers and then Famous Studios produced cartoons starring Superman.

the two reporters reveal the Nazis' location to American authorities, thereby ending the German threat in the area.

Richard Dyer has argued that "the cultural history of the past few centuries has been concerned with finding ways of making sense of the body, while disguising the fact that its predominant use has been as the labour of the majority in the interests of the few."[27] As a particularly wartime version of this history, the "Superman" cartoons exemplify the effort to represent the individual, male body—in this case, that of the superhero—as that which works for the welfare of everyone, and not while performing ordinary labor, but, rather, heroic deeds of global importance.

Despite this militarization of the male body and male behavior, the cartoons set in operation a sort of Newton's Third Law of Motion regarding representation. For all of the apparent sublimation of sexual desire in favor of patriotic, military necessity, there is nonetheless an equal and opposite intensification of sexuality, or, at least, female sexuality. *Jungle Drums* quickly transforms Lois Lane from star reporter to object of lust, as the film accepts the standard racist tropes of the danger white women face in black Africa. After the Germans capture her and she refuses to hand over secret documents, she herself is presented to the apparently all-male natives who tie her to a stake, build a fire, and dance around her. Then, in a doubly-determined happy ending, Superman not only crushes the Nazi ring but also rescues Lois, with the effort to win the war merging with the necessity to save North American white women from the proverbial fate worse than death.

Indeed, *Jungle Drums* may have been deemed fit for release without any censoring of Lois's predicament precisely because so much of the film is intertwined with the patriotic virtues. In addition, all sexual threats are implied rather than actual, from the dialogue ("I warn you, Frawline, unless you talk I vill make no effort to interfere mit dese natives") to the visuals of natives dancing while Lois burns.

Because it appeared after 1942, *Jungle Drums* needed to be analyzed by a member of the Library of Congress staff in charge of determining a film's historical value. This analysis, included with the copyright document (a transcript of the dialogue, from which the fractured German-English noted above has been taken), provides extraordinary evidence about the possibility for the wartime viewer, and particularly the female viewer, to interpret the film as stressing not the necessity for an Allied victory, but the physical and sexual threat to Lois. Unlike the description of *A Good Time for a Dime*, which implies a male viewer but provides no definite information, the Library of Congress staff analyses identified the analyst; *Jungle Drums* was viewed by Barbara Deming who, after her stint at the Library of Congress, became a poet, fiction writer, film reviewer (for *The Nation*), and activist for civil rights, women's rights, and nuclear disarmament.[28]

From reading her description of *Jungle Drums*, one can tell that Deming had watched a number of films for the Library of Congress, had become completely familiar with their brand of patriotism, and had grown more than a little tired of them. Describing the Germans trying to pry information out of the reporter from the *Daily Planet*, Deming adds an ironic comment: "But Lois is of course steadfast." Later, detailing Superman's rescue of Lois, Deming notes that the flames "are roaring pretty high by now," and early on she acknowledges the gaps in the narrative: "How they [the Germans] know the plane was carrying these papers I couldn't say."

Deming describes the threats against Lois in extreme detail, however, devoting far more time in her scene-by-scene analysis to these events than does the film itself, and using language that emphasizes danger. Deming writes that "Natives . . . have erected a stake, and are piling wood about it." She continues with, "Their native drums boom; they execute a frenzied hungry dance." Then, in one of the few such examples in the description, Deming cites a line of dialogue, underscoring the threat of sexual violence through her paraphrase of the German commander: "Unless you talk, we will not interfere with the natives!" Then she writes that "the pyre has been lit; flames lick up towards Lois; the natives prance around her."

The account of Lois's predicament, some 150 words, comes in a rather brief description—seven paragraphs and 650 words. Few other activities would be described so carefully, and none merited such double adjectives as "frenzied [and] hungry," as does the natives' dance around Lois. Moreover, given Deming's later commitment to civil rights, it is doubtful that her perception of the sexual danger was heightened by racist considerations; that is, that the menace came from black natives. Instead, the female spectator seems to have fixed upon the threat to the woman in the film, a threat which the film itself and much of the copyright documentation deemphasize in favor of other issues.

Deming's analysis points out the difference between the film's narrative strategy and its possible reception. In general, *Jungle Drums* seems completely in keeping with World War II Hollywood propaganda in which all issues become informed by the patriotic virtues, and also seems consistent with Paramount's desexualization of the cartoon star.

It may well have been that the obviously pro-Allied narrative exempted the film from the too-close scrutiny of the Hays Office. During the 1930s, a similar practice occurred in the office's review of quality pictures from the major studios. As a result, David O. Selznick's *Gone With the Wind,* or Ernst Lubitsch's *Desire,* could be cleared by the Production Code Administration because they were clearly excellent features, with sex made acceptable by the high gloss of studio production at its best.[29] Very probably, during the war, films that could make no real claim to the "Lubitsch touch" or to the Selznick imprimatur of excellence incorporated U.S. war aims as a means of blunting censorship. Deming's reading, however, indicates that despite the sublimation of sexuality that had taken place in Paramount cartoons, and despite the emphasis on wartime issues, sexuality itself, at least for certain audiences, was never far from the surface.

Controlling and Interpreting Cartoons

The article in *Look* made the censor's code regarding cartoon sexuality seem universally understood and absolutely determining not only of cartoon content, but also of an audience's interpretation of cartoons. Moreover, some film historians have tended to explain any seeming advances in the representation of sexuality, and particularly female sexuality, through an evolutionary model based on increased public enlightenment. In an otherwise insightful essay on Disney's *Peter Pan,* for instance, Donald Crafton interprets the 1953 film's handling of sensitive issues as indicative of an increased scientific and popular interest in sexual psychology.[30]

My analysis demonstrates that the representation of sexuality is never fully determined by censorship codes, and never simply and relentlessly progressing. Instead, the cartoon studios developed various strategies in relation to sexuality, strategies that were both intersecting and contradictory, and that were employed, for example, in the cartoons themselves, in the legal documents about the cartoons, and in the public relations material that described them in the popular media. For most of the cartoon studios, these strategies were influenced by Disney's dominance of animation in general and the child merchandise market in particular, and the concomitant need to differentiate their own product from that of the Master in order to maximize their audience. As a result, sexual relations could extend well beyond hand-holding, but even here practice did not divide simply between Disney and non-Disney animation, as Disney himself, for a brief period in the late 1940s, moved toward the Tex Avery mode of representation. In addition, as the documents in the Library of Congress demonstrate, different audiences may well have read films in different ways, some more censorable than others.

Thus, censorship, the law, the marketplace, and various segments of the audience vied for power over representational practice and interpretive strategy. The Hollywood cartoon studios themselves resisted their own strictures regarding sexuality in the different products that they made, products that constructed a hegemonic representation of sexuality and also a fractured, contradictory one.

2 | READING THE FILM BILL

Features, Cartoons, and the First-Run Theater

During Hollywood's classical period, most audiences saw cartoons and responded to them in movie theaters, where animated shorts combined on the film bill with live-action short subjects, newsreels, coming attractions, feature-length films, and, frequently, live acts, each one of them shown several times throughout the day, every day. The film bill probably developed from vaudeville—that is, a series of diverse segments which, when taken together, connote variety. But it also has links to wider industrial and social changes that had been taking place at least since the Civil War: a nationwide system of standardized time so that train schedules could be coordinated; the time clocks that came to govern urban factories and the concomitant demands by labor to shorten working hours; and even the development in other leisure, spectatorial pursuits of temporal precision (in boxing, for instance, the widespread adoption in the 1890s of three-minute rounds). Later, radio, probably the movies' chief competition prior to the 1950s, also specialized in precision and routine, playing the same programs every day or every week at exactly the same times. By the early twentieth century, then, travel and commerce, and also work and entertainment, had come to be defined, engaged in, and possibly even enjoyed to the degree that those activities could be measured and modulated by the clock.

THE FIRST-RUN THEATER FILM BILL

Precision became the point in the first-run film bill from the 1930s, the decade that I will be considering here. In October 1931, for instance, moviegoers in Washington, D.C., knew that *Wicked* (an Elissa Landi melodrama), played first at 11:18 A.M., and that the stage show began at 12:29.[1] Besides signifying the relentless dependability of movie entertainment, this time schedule maximized movie profits by allowing theaters to fit in as many shows as possible, and aided other businesses by letting patrons plan their day around transportation downtown, shopping in the area, and then going to the theater.

While reflecting the modern concern with time, the film bill also developed from changing perceptions of space. Movie entertainment exemplifies the same trend as museums, which grew in the eighteenth and nineteenth centuries from the private curiosity cabinets of the aristocratic and the wealthy to public palaces designed to educate the middle class. Museums provided the illusion of wholeness; patrons could see everything, from minerals and birds' nests to Egyptian mummies and Renaissance paintings. Similarly, with the movies, and particularly with first-run theaters, spectators attended modern, public spaces which seemed to present views of everything: various forms of fiction (including the cartoon), education (scientific short subjects, for instance, or travelogues), stage shows, and so on. Public entertainment brought together within a single space all spaces to be seen and appreciated.

Within this panorama of an afternoon's or evening's entertainment, the cartoon held a privileged place. Except in those rare cases when the feature film was itself a cartoon, the animated short stood out as the movie most unlike the bill's main attraction because of its obvious artificiality: the animals that spoke and danced, the ease with which objects changed shape or color, the painted rather than realistic mise-en-scène. Paradoxically, the cartoon was also, frequently, the supplemental film most *like* the feature. Just like the full-length, live-action film and different from the newsreel and many short subjects, the cartoon was a fictional narrative, depended upon a star system as a means of luring an audience, and made constant use of both comedy and performance (singing, dancing, etc.).

Within the context of the entire bill, the cartoon performed a function vital to the project of presenting the world to the audience. The importance on the bill of not only live-action fiction films, but also science films, travelogues, and other nonfiction genres, attests to the status of the live-action movie as that which could show practically everything. There were limits, however, and as a result, animation, with its apparently unbounded plasticity, became that which could depict even the unphotographable, those things that escaped even the live-action lens.

Thus, the commercial cinema developed a narrative of entertainment in which an individual film worked as part of a much larger system, one that included not only the entire bill but also, among other things, the popular journalism that brought discussions of movies to the public. Further, in making use of this narrative system, the cinema asserted the importance of the mode of representation (showing an entire bill of related films) rather than just the object of representation (star, story, mise-en-scène, editing, etc.). By studying the entire bill, and particularly the special relationship between cartoon and feature, we can begin to construct a history of going to the movies, and of what it must have been like to watch them.

In broad terms, the first-run film bill from the period can be divided into five major categories: 1) the bill that stressed the significance of the feature film by showing no supplemental movies, or just one—the newsreel; 2) the bill with a full array of films, all of which came from the same studio, and which, taken together, celebrated that studio as the producer of a wide range of entertainment; 3) the absolutely diverse bill, with differences from film to film usually developing around issues of class, genre, or culture; 4) film bills in which all or most of the movies were generically similar, or at least thematically so; 5) bills addressing a specialized audience—an adult one, perhaps, or a primarily female one. Many of these bills were apparently quite often consciously constructed to achieve certain effects; others can be read in spite of their seeming randomness.

These bills and this system of film exhibition helped teach audiences from the period about class, race, and gender and also about the difference between high culture and low, between adult and adolescent. Viewers learned, too, about entertainment itself, about what constituted its highest forms and also its simply acceptable forms. In short, first-run theatrical exhibition practice, through its strategies for regulating leisure, functioned as an agent of social control and as a means of transmitting specific cultural norms. The first-run patron saw him/herself represented in the films on the bill and in the combination of films;

but that first-run patron also learned how he/she ought to be represented, and thus, at least in part, how to make sense of the world.

METROPOLITAN ENTERTAINMENT IN THE 1930S

An analysis of Washington, D.C., first-run theaters from the 1930s helps us understand the interaction of cartoon and feature and also the experience of big-city film viewing. Information is available because, unlike most major newspapers, the *Washington Post* from this period arranged its reviews according to theaters. A typical page (fig. 9), usually run once a week, would include the names of all or most of the first-run theaters, followed by the title of the feature and its credits, a review, and then a mention of the supplemental bill, sometimes evaluative and sometimes simply in the manner of a list (the newspaper provided much sketchier information about subsequent-run, neighborhood theaters). This system of organizing reviews around theaters as opposed to individual films implies that at least during the 1930s, going to the movies meant something quite different from what it means now. At first-run theaters, audiences paid for a full menu of various kinds of entertainment, among which the feature was the principal but not the sole attraction.

By 1937, fifty theatres in Washington had been wired for sound films, and throughout the 1930s seven theaters, ranging in seating capacity from 1,500 to 2,700, dominated first-run exhibition: RKO-Keith's, Loew's-Palace, the Rialto, the Metropolitan, Loew's-Fox, Loew's-Columbia, and the Earle.[2] In addition, the Belasco showed an occasional art film (*Ecstasy*, which introduced Hedy Lamarr in a notorious nude swimming scene to American audiences, played there in 1936), and the Trans-Lux specialized in entire bills of short films. These were not the Depression-era theaters that used bingo and giveaways to increase attendance. Unlike the second- and third-run neighborhood houses that resorted to those and other tactics (double features, for instance) to bring in patrons, these theaters were elegantly designed and might cost between a quarter (typically a special reduced rate) to a dollar or a dollar and a half to attend.

Of the approximately 175 bills selected at random from these theaters

THE NEW CINEMA OFFERINGS

METROPOLITAN

LOEW'S FOX

RKO-KEITH'S

COLUMBIA

RECORD OF FERRARA LAUDED BY STIMSON

Retiring Cuban Ambassador Is Pan-American Board' Guest of Honor.

ABOUT THE SHOWSHOPS

With NELSON B. BELL

Figure 9. Typical *Washington Post* first-run movie theater listings from the 1930s (21 May 1932).

between October 1931 and June 1942, about half contained cartoons. With some variation, the first-run bill from the period consisted of a feature, a short subject, a cartoon, a newsreel, and a stage show of varying complexity and length (I am assuming that coming attractions also appeared on these bills, along with the occasional advertisement; the *Post* never mentioned these aspects of the bill). Affording closure to an afternoon or evening at the movies, the feature usually played last of all the movies on the bill.

The construction of the first-run film bill demonstrates the dominance and interdependence of just a few filmmaking companies during the period. Historians have divided the movie companies into the Big Five—Loew's, Inc. (which controlled MGM), Radio-Keith-Orpheum (RKO), Paramount, 20th Century–Fox, and Warner Bros.—and the Little Three—Columbia, Universal, and United Artists. The Big Five maintained their stranglehold on the market by owning a majority of first-run theaters in major urban areas, theaters that, though relatively few in number, generated the bulk of film profits. The major companies forced unaffiliated theaters (usually the subsequent-run houses) to buy films in blocks—that is, theaters could not simply rent one Warner Bros. film, but, rather, had to commit to several months' worth. This block-booking system applied to cartoons as well as to features. Exhibitors may have had to program an entire season's worth of MGM films in order to show one Clark Gable movie, and, similarly, a season's worth of Paramount cartoons for just a few "Betty Boop" shorts.

The first-run theaters in Washington tended to be affiliated with the major studios: Keith's with RKO, for instance, the Metropolitan with Warner Bros., and the Fox with 20th Century–Fox (and then, as the Loew's–Fox, co-owned by those two companies). Booking practice in these theaters was not quite so ruthless as with the unaffiliated ones. Regional managers did most of the scheduling, with help from other, more local managers (thus first-run theaters owned by the same company, but in two different areas, would schedule similar bills but with some differences according to local tastes).[3] A Loew's-owned theater probably would show most of MGM's output, but it would also present films made by other studios, and not necessarily rented as parts of larger blocks. As a result, the bills at first-run theaters were not so randomly generated as those at independent theaters. First-run theaters were relatively freer to construct bills along thematic lines, or in order to reach a specific audience.

Demographics

Washington's first-run theaters occupied an area downtown, between 12th and 15th Streets, Northwest, just above Pennsylvania Avenue. Along with an efficient bus system, at least three streetcar routes served the area. Nearness to major intersections made the area easily accessible by car, a not insignificant advantage when one considers that, in 1935, Washington had "the largest number of automobiles per capita of any city in the United States—more than one car to every three persons."[4] Moreover, this downtown location made it possible for residents of nearby Virginia to attend first-run movies in the District.

Those visitors from Alexandria or Arlington and moviegoers from Washington also could take advantage of all of the shops and department stores near the theaters, as the major shopping district ran between 7th and 15th Streets, north of Pennsylvania Avenue.[5] Hardly accidental, the proximity of theaters to stores fulfills the logic of late nineteenth- and early twentieth-century capitalism; all kinds of consumption would be extensions of each other, with the stores, restaurants, and theaters aiding each other's profits. The theater bills themselves often made variety their prime product, with the appearance of difference signaling the presence of a good buy. Thus the bills, as well as the location of the theaters, turned one of the bases of American mythology—pluralism—into practical economic theory.

In 1937, the District of Columbia had a population of 600,000, with some 200,000 more living nearby in the cities and suburban areas in Maryland and Virginia. Precise demographic information is difficult to come by, but, not surprisingly, one of the largest segments of this population worked for the federal government: in 1937, about 112,000, with roughly one-third of that number having moved to Washington with Roosevelt's New Deal expansion of the federal bureaucracy. Many of these workers and their families constituted the period's newly emerging, managerial middle class, with some statistical information supporting an educated guess that the city had more members of the middle and upper-middle class than most other urban areas. In 1935, for instance, District residents paid more in taxes than citizens of most cities, and among cities of similar size (including Minneapolis, New Orleans, and Cincinnati), Washington could claim the highest rental and home values and the most houses worth more than $5,000.[6]

In the midst of this affluence lived the fifth largest black population in the country, and, despite a sizable black middle class, most in this group did not have access to many of the benefits just mentioned. About 27 percent of Washington's population in 1937 was black, and as the Federal Writers' Project American Guide Series from that year maintains, "in some respects the Washington Negro receives better treatment, in others worse, than his brother in the 'deep South': for example, he may occupy any seat of a public vehicle, but no seat whatever in any theater other than those maintained for his own race." Forty percent of the black population was unemployed, and in 1935 black recipients accounted for more than 70 percent of District relief cases.[7]

Almost certainly segregated, then, whether by law or by custom (in the above citation, it is uncertain whether the Guide is referring to movie theaters as well as to legitimate theaters) the first-run theater in Washington benefited from a relatively affluent clientele. But like other successful forms of entertainment from the nineteenth and twentieth centuries—the English music hall, for instance, or American vaudeville—the theaters must have sought a broad-based audience in terms of class (albeit with an emphasis on the middle class), if not in terms of race.[8]

Celebrating the Studio

Keeping this audience overview in mind, we can begin to examine the bills themselves. For a very few theatrical bills, a single film, playing by itself or with only modest support, and certainly not with a cartoon, signified the highest class of entertainment. In 1932, for example, MGM's *Grand Hotel* opened with no supplemental films at the Columbia Theatre, and with all seats reserved for the two daily performances (as opposed to the more common five to seven) and the three on Sundays. The film itself, at 113 minutes, ran longer than the typical Hollywood product from the period, but that length alone would not rule out a supporting bill. Just four years later, for instance, *Mr. Deeds Goes to Town*, at 115 minutes, played with a "Betty Boop" cartoon and a newsreel. In the case of *Grand Hotel* the perceived quality of the film made it stand above the typical Hollywood product, and therefore above the typical mode of exhibition, with its full range of movies. At least two other MGM films from this period, *Rasputin and the Empress* (1933), the only sound film featuring all three of the Barrymore siblings, and 1936's Oscar-winner for Best Picture, *The Great Ziegfeld*, appeared with newsreels only.

Playing with the feature, the newsreel might seem to stand outside entertainment and signify, instead, a commitment to news, to real life even in the midst of Hollywood make-believe. In part because of its position outside entertainment, the newsreel became the most indispensable portion of the auxiliary bill. The cartoon, or the stage show, or the short subject could, because of the feature, seem to be simply redundant. The news, however, attested to the importance of going to the movies for something other than mere entertainment.

Another strategy stressed variety and abundance rather than singularity. On October 30, 1931, the Metropolitan, a new theater controlled by Warner Bros., opened in Washington. The next day, the *Post* noted that "last evening's inaugural bill embraced a salutary address by Mr. Albert Howson, chief of the Warner Brothers' scenario department, and a few brief remarks by Brig. Gen. Pelham D. Glasford, recently appointed Chief of the Metropolitan Police." The review mentioned that the bill supporting the feature, *Svengali*, included a newsreel, a Robert L. Ripley *Believe It Or Not* short, a "Merrie Melodies" cartoon, and *The Road to Mandalay*, a travelogue about a trip from Sumatra to Burma. The evening concluded with a "special reel showing the guests of the evening entering the theater to enjoy an exceptional bill."[9]

Here the bill of Warner Bros. films clearly represents the amazing diversity of the studio's product (Warners produced the feature, the Ripley short, the cartoon, and very probably the travelogue), while the presentation and the subject matter of the films signify the studio's overall relationship to all manner of authority. Even before the films began the audience witnessed the equation of the studio with the national and local patriotic virtues, as the scenario "chief" shared the stage with a retired general who now served as police chief. The feature film starred John Barrymore as the hypnotic Svengali, while the Ripley short and the travelogue stressed the studio-as-educator. Then, the special reel demonstrated Warner Bros.' technological mastery along with the cinema's ability not simply to amuse its spectators but, through the camera that had been trained upon them, to turn surveillance into entertainment. At the same time, those in the audience could watch films, and also, because of the special reel depicting that audience entering the theatre, themselves enter the spectacle.

In this context, the cartoon, probably *You Don't Know What You're Doin'*, which was released on October 21, functioned as unadulterated entertainment, and therefore as a further indication of the variety of the Warner Bros.' product. Besides standing in generic contrast to the feature and to the live-action shorts, the cartoon's title song was almost certainly owned and marketed as sheet music by Warner Bros., and the cartoon itself includes a reference to Al Jolson, who at the time was under contract to the studio. From this, we can begin to see the complexity of first-run exhibition. A studio could be inscribed and reinscribed within a bill, both from film to film, and, as with the song and the Jolson reference, within a film. The entire bill sold the studio's product, because as each film began with the unmistakable Warner Bros. logo, an audience could not help but be aware of the source of all of the movies. As a result, the film bill sold the audience on returning to a theater that was owned by Warner Bros., and, in the case of the cartoon, sold other Warner Bros. products, here the song and a star.

Two years later, Warner Bros. employed a similar strategy to introduce new, lower rates to the Metropolitan. The *Post* reported that "the new policy embraces a two-and-a-half-hour show with a first-run screen attraction and a surrounding program of short reels," with all admissions reduced to 25 cents before five P.M., and a 25-cent balcony price and 40-cent orchestra admission from five until closing. "The first attraction for the new policy," the *Post* reported, would be *Dark Hazard*, with Edward G. Robinson, playing with "short-reel featurettes . . . headed by 15 minutes of sound news," and including a live-action parody of MGM's *Dinner at Eight*, a "Merrie Melody" cartoon starring Buddy and Towser, and a "Vitaphone novelty" featuring the Mills Blue Rhythm Band.[10]

On this bill, the very availability of a wide range of film genres at prices that would allow anyone to sit in the orchestra signifies quality entertainment. Thus the thinking at Warner Bros. stood in opposition to that at MGM. *Grand Hotel*, playing by itself and with all seats reserved (and presumably at reserved-seat prices) came to equate the experience of going to the movies with that of more elitist activities, such as attending the legitimate theater. Warner Bros., however, with a melodramatic feature film (about a compulsive gambler), a parody of another studio's product, a newsreel, a comedy cartoon with music, and a musical short, sold volume at bargain prices. Indeed, this cursory look at theatrical bills that were special—glossy features, openings of new theaters, an introduction of a new pricing policy—would seem to substantiate the familiar thinking about Warner Bros. during the early 1930s, and also about MGM: the former, through subject matter and exhibition strategy, stressed its connection to the working class, while the latter, once again through film subject and exhibition, catered to a more middle-class clientele.

Cartoons and Classics

Another bill from 1936 that clearly was aimed at a far narrower, primarily adult audience used a cartoon in a central role, thereby helping to demonstrate how some evenings at the movies were constructed to connote high quality. In May, *Ecstasy*, the film that made Hedy Lamarr a star, opened in Washington after already having been granted the dual status of art film and *cause célèbre*. The *Post* reviewer noted in the opening paragraph that the Czechoslovakian film had been the "winner of several prizes for cinematographic art," then referred to it as an "art picture," and to the "public curiosity piqued by the picture's hectic history on this side of the Atlantic." Then, further indicating that this was an adult film, the reviewer called *Ecstasy* "psychoanalytic."[11]

Because the theater, the Trans-Lux, apparently was unaffiliated with any major film company, the management had the responsibility for constructing the bill. And because the film industry stressed general appeal in order to maximize box office, the exhibitor faced the problem of creating specialized and profitable adult entertainment within the limits of block-booking practice. As a supplementary bill, therefore, the theater offered "a New Orleans pictorial, Mendelssohn's 'Fingal's Cave,' with colored nature pictures, the Paris Symphony Orchestra playing 'Der Freischütz,' and a colored cartoon of Don Quixote in his padded cell, before and after the windmill incident."[12]

Despite the generic diversity—travel, nature, classical music, cartoon—the entire bill signified mature entertainment, and the selection of the cartoon is particularly interesting. *Don Quixote* stands out as one of Ub Iwerks's more unusual films from the period. Many of the seventy-five cartoons that he produced after leaving Walt Disney and running his own studio from 1931 to 1936 were taken from nursery rhymes or fairy tales: *Jack and the Beanstalk*, *Puss in Boots*, *The Brave Tin Soldier*, *The Valiant Tailor*, *Simple Simon*, *Humpty Dumpty*, *Little Boy Blue*, *Dick Whittington's Cat*, *Bremen Town Musicians*, *Mary's Little Lamb*. *Don Quixote*, however, was his only cartoon based on an acknowledged literacy masterpiece. Further, although it parodies Cervantes, the film itself is not fully parodic; it concerns a contemporary character who is crazy and has been placed in an insane asylum (an atypical protagonist and setting for a cartoon), who imagines himself to be Cervantes' hero. In this appeal to a knowledge of adult fiction, the cartoon, along with the Mendelssohn and the Paris Symphony shorts, worked to confirm the feature film's status as different from, and more mature than, the typical Hollywood product.

The undeniable high culture of the entire supplemental bill, from the cartoon to the classical music, underscored the quality of the feature, thereby offseting its potential as pornography (Lamarr's nude scene), and so helping relieve the theater of a possible problem in public relations.

In the case of *Ecstasy*, the *Post*'s insistence on its psychological and aesthetic value contributed to an adult discourse about the film that was matched by the bill's supplemental movies. Yet another bill, constructing a similar kind of quality entertainment and demonstrating the same intersection of motion pictures, journalism, and reputation, shows that the *Ecstasy* bill at the Trans-Lux was part of a general strategy of first-run exhibition. A few months after Lamarr's film opened in Washington, Paramount reissued a movie from 1930, *Morocco*. When that film showed in Washington, it played with reissues of two cartoons, Walt Disney's *Three Little Pigs* (1933) and *The Big Bad Wolf* (1934). Indeed, the reviewer began by speaking of the Disney cartoons, saying that "a great picture and its equally renowned sequel for the first time are being shown on the same bill at the Rialto Theater." After naming the two short films, he called them "mighty in sheer entertainment," and asserted that "this return engagement of the pinkish porkers is of moment only slightly less than the presentation of *Morocco*, the picture which presented Marlene Dietrich in her first production of Hollywood origin." Finally speaking solely about the feature presentation, the reviewer concluded that "the passage of six years . . . has not dulled the worth of the performances of [Gary] Cooper and Miss Dietrich."[13]

We tend to think of *Morocco* either as an auteur masterpiece rediscovered in the 1960s, or as a cult, camp classic famous for Dietrich's cross-dressing and for its last scene, in which the star kicks off her pumps and follows her French Legionnaire lover into the desert. The review shows, however, that even in the 1930s the film was considered an important one, and not least because it was Dietrich's first American film. But the Disney films were believed to be just as important—indeed, the review calls them "great" and "renowned"—and just as worthy of a fresh viewing. The usual bill hierarchy remains constant, as the cartoons must be of "slightly less" moment than the reissue of the feature. Given that hierarchy, however, there is also a surprising equation, on the one hand of Dietrich/pigs, and on the other *Morocco*/Disney cartoons, with each aspect of each equation signifying the importance of the other, and all of them combining to guarantee the significance of the entire bill. Very possibly for the first time, a review categorized a cartoon as a classic (and even as an historical artifact; that is, worthy of a reissue), and acknowledged animation as a means of selling an even greater artistic triumph, the feature film.

Managing Difference

Occasionally, the theatrical bill stressed the quality of a single film, or of a single studio, or of a program of "classic" films. Generally, however, the bill seemed governed either by a desire for absolute genre diversity or for genre sameness, the former seemingly for the widest possible clientele, the latter for any of various more specific audiences.

Neither of these two types of film bill gained the unproblematic critical acceptance that accompanied such special theatrical events as the first-run showings of *Grand Hotel* and *The Great Ziegfeld*. In the *Post* on October 3, 1931, for instance, movie critic Nelson Bell reviewed both *Monkey Business* and *Devotion*.[14] The former, with the Marx Brothers, played with a "Scrappy" cartoon, *The Little Pest*, which Bell found "amusing with its flirtatious fishing worms showing us just why fishes do bite." *Travelaughs to Reno* accompanied the film, as did a film that sounds at least ostensibly educational, *Pearls and Devil Fish*. Bell, however, found the comedy, from the Marx Brothers to Scrappy to Reno, a little too relentless, and complained that "someone with rare foresight should bring on a tragedy."

Never easy to please, Bell objected to too much diversity on the other bill. A "romantic comedy-drama of middle-class English life" starring Ann Harding and Leslie Howard, *Devotion* appeared with a "Tom and Jerry" cartoon, *Wot a Night*. Not the cat and mouse Tom and Jerry (who didn't appear in cartoons until 1940), this was a pair of cartoon stars that enjoyed some popularity during the early 1930s. Similar to Mutt and Jeff in appearance, both Tom and Jerry were clearly working class, and, as the title indicates, the speech used in their films was as far removed as possible from "middle-class English," but conformed to the stereotypes of poor and uneducated speech. *Julius Sizzer* also appeared on the bill, an "obvious travesty" of Warner Bros.' recently released gangster film, *Little Caesar*, and starring Benny Rubin "in a dual dialect role." The bill, then, ran from the cultivated feature, to a parody short, the title of which poked fun at high-brow culture, to a cartoon that concentrated on the lower class.

From this, we can start to guess at the complex demographics of even the first-run theater audience, if we assume that, at least to some degree, the films on the bill reflected the spectators watching them. Or, the bill demonstrates the ways in which an entire theatrical experience represented to middle-class patrons, who most consistently could afford first-run prices, not only their own class, but also the lower class. Bell, however, invoked principles of aesthetic wholeness and complained that the audience, whatever its makeup, was not treated to a more generically consistent evening: "The supporting program," he wrote, "while diverting enough in a cruder and less finished way, is not at all times commensurate with the quality and the artistry of the major feature."

Despite Bell's complaint's about inconsistent bills, many theaters in the *Post* sample adopted a strategy of generic difference between feature, cartoon, and other supplemental shorts. Bills display the strong connection of 1930s first-run exhibition with vaudeville, with its animal acts and opera divas sharing the stage. In 1931, *Consolation Marriage*, a Barbara Stanwyck melodrama about mismatched couples and unrequited love, played not only with a variety stage show, but with a "Krazy Kat" cartoon called *Weenie Roast*. When Fredric March and Miriam Hopkins appeared in *All of Me* in 1934, they were supported not only by a Charley Chase comedy, but by a "Popeye" cartoon (probably *Let's You and Him Fight*). In this last example, a romantic leading man—March—shares billing with a decidedly unromantic one—Chase. Furthermore, the feature, about an upper-class woman who learns a valuable lesson from a slum dweller, finds itself on a bill with a cartoon that almost certainly, in the manner of "Popeye" cartoons from the period, unsentimentally depicted comic, ill-spoken, ready-to-roughhouse working-class characters. Similarly, *Women of Glamour*, with Melvyn Douglas as a wealthy modern artist, opened in March 1937 with another "Popeye," *Organ Grinder's Swing*, once again asserting class diversity on a typical bill, along with the various possibilities for masculine behavior.

The diverse theatrical bill was marked not simply by generic difference, but by class difference both implied and actual, indicating the manner in which film exhibition controlled class tension through an expression of that difference. For instance, a 1936 bill featuring *Next Time We Love*, with James Stewart and Margaret Sullavan, played with a newsreel of King George V's funeral and a cartoon about the Toonerville Trolley and Katrinka (probably the Van Beuren studio's *Toonerville Trolley*). In this case, a love-triangle melodrama, perhaps the Hollywood genre par excellence because of its roots in the nineteenth-century novel and its connection to the rise of a bourgeois audience, shared the screen with an elegy for the ruling class (the newsreel) and a knockabout comedy about the working class (the cartoon). When *Lloyds of London* opened in Washington in 1937, the celebration of the British mercantile class had, as a supplementary film, a Terrytoon animated short featuring Farmer Al Falfa (probably in *Barnyard Amateurs*), a star who would make films for only another year or so, in part because he represented a rural, farming class soon superseded in American mythology by the more urban, entrepreneurial heroes of the feature. *Fire Over England*, a 1937 British historical drama celebrating not only the defeat of the Spanish Armada but also the political acumen of Queen Elizabeth, played with Walter Lantz's low-budget *Duck Hunt*, starring Oswald, a sort of rabbit-of-the-people.

Other bills marked a sliding scale of culture as well as class. When *Intermezzo* opened in Washington, it played with a live stage show and a Terrytoon cartoon (probably Gandy Goose in *Hitchhiker*). The feature possessed all the markers of Hollywood high class and high culture: it was produced by David O. Selznick, and starred a European discovery in her American debut (Ingrid Bergman) along with an established British actor known for his portrayals of tormented artists and aristocratic gentlemen (Leslie Howard). The stage show combined high culture and low within just one act. The headliner, Larry Adler, played the harmonica, perhaps the quintessential working-class musical instrument, but he used it to perform the classics (in this appearance he played Brahms's Second Hungarian Rhapsody). There were also some less problematic acts on the bill, that is, acts that were more identifiably popular—a singer, a puppeteer, and a band honoring the local football team with "Redskin Romp." Gandy Goose, the star of the cartoon, was modeled after vaudeville comic Ed Wynn, with this likeness clearly aligning the animated short with an extremely popular form of entertainment and an extremely popular entertainer.[15]

In 1937, when Katharine Hepburn appeared in *Quality Street*, an adaptation of James Barrie's play, she shared the bill not only with an "Oswald" cartoon, *The Birthday Party*, but also with a "March of Time" which the *Post* described as recounting "the rise of voodooism in Harlem . . . and the crisis, as reflected in Lloyds . . . occasioned by the abdication of Edward VIII." Depicting an exoticized, nonwhite, lower class and also a trauma affecting England's ruling class, then, the newsreel acted as a balance between a delicate, serious feature and the more infantalized, comic cartoon.[16]

The narrative of an evening at the movies seemed complete because it so frequently offered the range of human possibility (at least to a white audience), from Fredric March to Popeye the Sailor. As in the case of *Quality Street*, moreover, the bill also seemed perfectly balanced, from Katharine Hepburn to Oswald the Rabbit, with a newsreel, about Edward VIII and witch doctors in Harlem, that incorporated elements associated with both of the evening's stars.

In depicting the working class, the cartoons on these bills, no matter how problematic their class ideology, exemplify a typical representational strategy for the animated films of the period. From Popeye to Betty Boop to Mickey Mouse to the original Tom and Jerry, cartoon stars, probably with more frequency than their live-action counterparts, tended to portray characters who were quite clearly poor. As a result, even when a film bill highlighted a feature about the upper class, there generally was a place reserved for the lower class. This supplemental part of the bill may have functioned as a safe place for expressing lower-class discontent—Mickey Mouse (in *Moving Day* [1938], for instance) might battle a sheriff trying to evict him, or Donald Duck recoil from his construction site boss (in *The Riveter* [1940]). But this frequently lower-class portion of the bill also served to commodify class in general. The spectator could go to a first-run theater and feel that he/she was getting a good buy for the admission price, with the class diversity on the entire bill coming to equal entertainment variety.

An equation between diversity and value also functioned in relation to race. On March 9, 1934, the *Post* announced that the RKO-Keith's theater would be holding over *It Happened One Night* for a third week, a long run for the period. Apparently, the theater management believed that the cartoon on the bill contributed at least minimally to the feature's success, as the *Post* added that "the second in the series of 'Amos 'n' Andy' cartoons [*The Lion Tamer*], wherein the actual voices of the two famous radio personages are recorded, is also being continued into the third week."[17] Produced by the Van Beuren studio, the "Amos 'n' Andy" series consisted of only two cartoons, and, according to Leonard Maltin, "neither [of the films] made the impact everyone thought they would, and the series came to an abrupt halt."[18]

This single instance from Washington may belie Maltin's assertion, or the cartoon may have benefited simply from being paired with one of the most successful feature films of the year. It is also possible that the two films were so ideally paired ideologically that the theater management knew instinctively to keep them together. *It Happened One Night*, of course, presents a Depression-era fantasy primarily about a lower class in which black people have no place; all of the bus passengers are linked by their poverty (except for the runaway heiress), and also by their race. The film interprets the Depression as affecting almost everyone on the one hand, and only white people on the other. The cartoon, however, solely depicts an unintelligent black lower class and the comic efforts of its members to make money. As a result, the theater provided a separate but equal entertainment bill. Because of the obvious differences between the feature and the cartoon, and because the cartoon incorporated well-known characters from a famous radio program, the audience clearly was being given value. But under the guise of racial diversity and entertainment variety, the audience was presented with a Depression untouched by racial tension. The bill provided two views of the Depression in which different races could be kept conveniently separate, both in the film and, probably, in the theater audience.

It could be argued that this extreme diversity concerning not only genre, but also class, race, and culture, indicates that the first-run cinema during the 1930s functioned in much the same way as the storefront theater and nickleodeon in the early part of the century. That is, it was a democratic space, one that offered something for everyone, a kind of entertainment utopia described by such critics as James Agree.[19] But historians have come to reinterpret the early theatrical experience, and to reassess, in the words of Miriam Hansen, precisely "whose public sphere" the theater constituted.[20]

If, as Hansen and others argue, the nickleodeon marked on the one hand the beginning of a readily accessible, always available bourgeois public sphere, and, on the other, the location where the materials of human existence became commodified, then the theater of the 1930s marked the full development of just such a space. This means that the cinema bill managed diversity by asserting it; in other words, differences—of race and class, and also different kinds of culture—could be made to disappear, replaced by a sense of aesthetic wholeness. Or, if not vanish altogether, these conflicts could be seen not to be conflicts at all. Both the theater and the film bill worked as a reaffirmation of liberal American principles of competition, difference, and similarity, all in peaceful coexistence, and thus part of the mythology of pluralism. Difference came to seem democratic, as the bill provided everything.

Diversity may have functioned not only ideologicaly, but also commercially. The clearly subsidiary status of the cartoon served to highlight the specialness of the feature and to build audience support for it. For example, when, in 1931, a Walt Disney "Silly Symphony" played with MGM's *Cuban Love Song*, it may have indicated an attempt by the film company to create a favorable response to the star of the feature, opera singer (and radio and recording industry star) Lawrence Tibbett, whom Hollywood tried for several years to turn into a major movie celebrity. To the film-going public, Disney probably enjoyed more renown than Tibbett, and even as early as 1931 his cartoons benefited from a critical discourse that stressed their artistry. Similarly, in 1937, Popeye, an established star in both comic strips and cartoons, appeared in *I Never Changes My Altitude* in support of *Mr. Dodd Takes the Air*. The latter starred Kenny Baker, who had broken into films the year before but never matched in that medium the fame he had enjoyed as a band singer. Showing the ways in which ostensibly competitive film companies worked to sell each other's products, the Paramount "Popeye" cartoon, playing before the feature, may have made audiences feel more comfortable with Baker, the Warner Bros. star whom they might have been seeing for the first time.

From this, we can see the marks of the Hollywood oligopoly, or control by just a few companies. Just as the typical bill highlighted diversity but also managed it in order to smooth over social tensions, so too did the major studios assert a belief in competition while frequently practicing collusion. Rather than insisting on individuality, the studios worked together to establish a kind of group uniqueness. Douglas Gomery has demonstrated that it was through their ownership of first-run theaters that the Big Five "operated as a collusive unit, protecting each other, shutting out all potential competitors, and guaranteeing profits for even the worst performer, usually RKO."[21] Although MGM, for example, was deeply concerned with its own profits, the studio also had a stake in the success of the other major studios as a means of preventing new competition from arising. The strategies for combining films from different studios on the first-run bill was part of this practice. While prioritizing the feature films of the single studio that owned it, a first-run theater often served to celebrate the product of all of the major studios. An evening at the movies proved beyond a doubt that a few companies (which placed their own distinctive, easily recognizable logos at the beginning of all their films) could meet all of an audience's needs for entertainment. Thus, while stressing competition with each other—cultivating their own stars, genres, and technological advances, for example—the studios, as exemplified by the film bill, also worked actively for cooperation.

Managing Sameness and Constructing Audience

Some bills are marked by such difference that they seem to defy any theorizing on the functions of diversity. In December 1939, for instance, the historical romance *Drums Along the Mohawk* played with a Robert Benchley comedy short and the Terrytoon *Wicky Wacky Romance*. But this is not to say that theaters and film companies always concentrated on creating the most diverse bills possible. Indeed, from the *Post* sample, a significant number of theatrical bills exhibit an apparently determined effort at similarity between the feature, the cartoon, and the other films on the supplemental bill.

On August 6, 1932, for example, the *Post* reviewed *The Man Called Back*, a melodrama set in England and the South Seas, which played with *Ghingi*, a travelogue about Australia and "its amazing animals and strange people," and an "Aesop Fable" cartoon set in China. In this case the theater, RKO-Keith's, provided a program about travel, ranging from the exotic (the feature) to the educational (the travelogue) to the comic (the cartoon). And the bill was not merely a function of studio convenience, of a theater playing the films provided by the producing company affiliated with it. Though RKO released the "Aesop" cartoons, the feature was an independent release, coming from the Quadruple Picture Corporation. Apparently, this was a bill very consciously constructed around a specific genre.[22]

The bill with *The Man Called Back* did not necessarily represent part of a pattern, an insistence that a travel feature must be part of a program about travel. Just a few months earlier, for instance, in May 1932, *The Blonde Captive*, a "pictorial record of the Northwest Australian Expedition Syndicate's explorations in the Antipodes," played with a comedy short and a cat versus mice "Merrie Melodies" cartoon, *It's Got Me Covered*.[23] Travel thus became just one part of a night's entertainment, rather than the theme for an entire bill. Indeed, the very lack of a pattern is interesting here. We tend to think of the Hollywood cinema from the period in monolithic terms: a vertically integrated industry, rigidly controlled production, and elaborate regional and national systems of distribution and exhibition. Once the industry's films were in the first-run theaters, however, the mode of presentation of those films, while conforming to certain strategies developed by national, regional, and local managers, and certainly not devoid of ideological functions, was, relatively speaking, up for grabs.

Understandably, the feature films that tended to share a bill with similar cartoons were comedies and musicals, as Hollywood animation itself tended toward humor and performance. In 1937, for example, MGM's *Broadway Melody of 1938*, with Eleanor Powell, Judy Garland, and Sophie Tucker (and also the decidedly unmusical Robert Taylor), and which the *Post* reviewer called a "revusical," shared a bill with *Mickey's Amateurs*, a Disney cartoon that was itself a series of performances. The latter bill points toward the importance of the Disney product. A short called *Rocky Mountain Grandeur* played with the two films, as did a newsreel notable for its depiction of "the Chinese situation." The theater, then, provided something of a split bill, two films about performance and two largely about travel. That a feature film should be aligned so strongly, in generic terms, with a cartoon, and set off just as strongly from the rest of the supplementary bill, probably demonstrates the perception that a Disney cartoon could be used to help sell an audience on the feature presentation. Mickey Mouse was a bigger movie star than anyone in *Broadway Melody*, and MGM may have counted on that star's presence in his musical boosting the approval of Powell and Taylor in theirs.[24]

Despite the generic sameness between comedy feature films and so many Hollywood cartoons, the presence on the bill of one did not require the other. *Palmy Days*, with Eddie Cantor, had no cartoon support when it opened in Washington in October 1931 (it played with a *Football Thrills* short, an installment of the "Paramount Sound News," and a "Barkville" all-dog comedy). Nor did *Million Dollar Legs* (1932), starring W. C. Fields, play with a cartoon. Perhaps Cantor and Fields, both with large followings from their vaudeville careers and both specializing in verbal humor, may have appealed more to adults than to children, and as a result booking managers may have been less prone to supplement their films with cartoons.[25]

The sample from the *Post*, however, indicates that, at least in first-run theaters, cartoons had not yet been classified primarily as children's entertainment. In May 1936, Shirley Temple's *Captain January*, a typical comedy/drama/musical celebration of the prodigy's talents, played with a musical short subject, a comedy short, and the "Metrotone News," but no cartoon. Nor did *Wee Willie Winkie*, in 1937, appear with a cartoon. In October 1936, however, another Temple film, *Dimples*, did indeed play with a cartoon, a "Mickey Mouse" called *Alpine Climbers*. This cartoon-optional approach to Temple's movies, which, although they had a large adult following, were also produced as children's films, indicates that scheduling a cartoon was not always an aspect of a film company's strategy to please the youthful section of its clientele. Even when *Captain*

January moved to the residential, second-run theaters, where, because of the lower prices and the proximity to people's homes, children undoubtedly made up a larger percentage of the audience than in the first-run theaters, apparently only two of the five theaters showing the film (and advertising the supporting bill) supplemented it with a cartoon.[26]

The Gendered Bill

The next chapter examines the logic of military propaganda, which leads to a mode of address directed primarily at a male audience. The logic of the marketplace, however, requires a different system: an address to a mass audience and, within that largely undifferentiated mass, several specific ones. As a result, first-run movie theaters in Washington maximized profits by attracting a wide clientele (within the boundaries of the District's legal and de facto segregation) and also by showing a commitment to the perceived special interests of some smaller groups, with a special emphasis on gender.

In May 1932, the feature *So Big*, starring Barbara Stanwyck and based on Edna Ferber's homage to maternal sacrifice, played with a "Betty Boop Talkartoon," *A-Hunting We Will Go* (along with a newsreel and a "sepia revue").[27] I do not mean to argue here that Betty's appeal, or Stanwyck's for that matter, was ever strictly female. At this time, Stanwyck specialized both in women's films like *Consolation Marriage* (1931) and *Shopworn* (1932), and in movies in which she was presented as obviously sexy, and therefore appealing to a male audience—*Baby Face* (1933), for example. As a result, her appearance in one kind of film might well attract a fair number of fans who enjoyed her in the other. Moreover, as chapter 1 indicates, the Fleischer Studio often made Betty's sexual come-on as overt and male-directed as possible. Still, the pairing of Stanwyck and Betty could not have been purely accidental. The Metropolitan Theatre, which showed the films, was aligned with Warner Bros., and *So Big* itself had been produced by that studio. Like all other "Betty Boop" films, however, *A-Hunting We Will Go* was distributed by Paramount. Thus, a bill combining the most significant female cartoon star of the period with a movie clearly designed as a woman's film, and based on a book by one of the period's most celebrated writers who also happened to be a woman, did not simply present one studio's product. This implies that, while never giving up on attracting the broadest possible audience (a movie from a Ferber novel certainly would be popular among men) film companies also engaged in demographic projects designed to focus on specific groups of viewers.

Another bill further demonstrates the point. In 1937, MGM produced *The Last of Mrs. Cheyney*, a Joan Crawford vehicle that changed a story about a male jewel thief—the ever-popular and frequently filmed *Raffles*—into one about a woman. In Washington, the film played with Disney's *Mother Pluto*, which the *Post* review described as "the famed Disney hound mothering a brood of chicks," and also a newsreel and a "What Do You Think?" short about telepathy.[28] The bill provided two films about gender switches, an arguably feminist one in which a woman gets to play a traditionally male role, and one in which a male character—Pluto—displays his maternal instincts. In addition, the live-action short stressed a feminine subject, as film companies in the 1930s used telepathy and the related subjects of astrology and the occult as plot devices in a number of women's films, or as the pseudo-scientific interests shared by any number of mindless upper-class women depicted in the period's comedies.[29]

Some bills seem designed for a male audience. In October 1936, *Texas Rangers*, starring Fred MacMurray and Jack Oakie, played with a "Popeye" cartoon, *Little Swee' Pea*, and also some newsreels. Thus, a western, perhaps the male genre par excellence, appeared with a cartoon starring the period's most famous sailor man. But one of the more interesting aspects of a study of theater bills from the 1930s is not so much their appeal to a masculine audience as their construction of masculinity. Hollywood animation was relentlessly male, from production practice (very few women held mid- or high-level jobs at animation studios) to the productions themselves (there were very few female cartoon stars). As a result, a cartoon from the period, invariably starring Mickey or Popeye or Oswald or Bosko or Tom and Jerry, could lock in the signification of a bill already preponderantly male.

Mickey Mouse functions as an interesting example of this effect. He appears more than any other cartoon star in the bills from the *Post* sample—eight times. As a comparison, Popeye supplemented six features and Betty Boop only three. In analyzing those bills, I am aware that my readings may demonstrate a certain retrospective awareness of Mickey's importance. While Mickey was acknowledged as a major popular icon in the 1930s, he has, in the decades that followed, become not only a monument of United States culture but of transnational capitalism itself.[30] Nevertheless, Mickey Mouse's presence on the theatrical bill proves instructive, as the Disney character came to be strongly linked to notions of masculine heroism and, ultimately, high-class entertainment.

Early in the period covered by the *Post* sample, Mickey took part in programs that, through their male characters, mythologized stereotypical American values, both contemporary and historical. In October 1931, for instance, Mickey, in *Fishin' Around*, supplemented *The Spirit of Notre Dame*. The feature itself opens with a dedication to the patron saint of Notre Dame football, the recently deceased Knute Rockne. Then, the film details the story of two Notre Dame athletes who, through football, overcome their personal animosity. A detective short based on *The Cat's Paw* also supplemented the feature, as did a tirelessly collegiate orchestra that played "Sweetheart of Sigma Chi" and "Washington and Lee Swing."[31] On a bill suffused with spirit, Mickey became part of a Depression-era narrative of an afternoon or evening at the movies that, in the feature, solved hardship through Catholic, all-male education and physical courage, provided the possibility for absolute truth through the male detective in the short, and, if the cartoon's title is an indication, depicted an abundance of leisure time for typically male activities.

A week after appearing opposite Knute Rockne, Mickey supplemented *The Cisco Kid*, in which Warner Baxter played a lovable Mexican bad guy engaged in various businesses, legal and otherwise, "on our side of the Rio Grande."[32] Here, Mickey (and it is difficult to determine in which cartoon) became the modern, homegrown, everyman counterpart to the Kid-as-nineteenth-century exotic. The bill projected an idealized historical progression from Wild West heroism, in which Americanness—as in entrepreneurial capitalism enforced not so much by law as by a revolver—could still be defined by foreignness, to a more practical, fully assimilated value system signified by Mickey. Going from romance to comedy, the bill nevertheless posited both heroes as typical, but with Mickey higher on the evolutionary scale than the Kid in signifying just what it means to be an American, or, more precisely, an American man.

Indicating Mickey's growing importance as an aesthetic object rather than the mouse-next-door, by the late 1930s he came to represent high-class, quality entertainment, the unique rather than the typical. In 1936, Mickey in *Three Blonde Mouseketeers* appeared on the bill with Walter Huston in *Dodsworth* and Robert Benchley in a comedy short, *How to Vote*. That same year, *Mickey's Grand Opera* supplemented *These Three*, and then, a year later, Mickey's *Moose Hunters* shared the screen with *The Good Earth*. In each case Mickey accompanied an adaptation of a work of literature that had achieved the status of a classic: novels by Sinclair Lewis and Pearl Buck (who won a Pulitzer Prize for hers), and a play by Lillian Hellman. Further, *These Three* was produced by Samuel

Goldwyn, an independent producer known for his quality pictures, while the other two featured acknowledged great actors: Paul Muni and Luise Rainer in *The Good Earth*, and in *Dodsworth*, Walter Huston.

By this time, a "Mickey Mouse" cartoon had become the most appropriate first-run supplement to films marketed as important and adult. In addition, Mickey had come to serve as the cartoon star equivalent to a range of revered popular icons. On the same day that the *Post* reviewed *The Good Earth*, the newspaper also reviewed the short-film-only bill at the Trans-Lux.[33] Mickey appeared opposite newsreels showing not only such domestic idols as Henry Ford, Joe DiMaggio, and Lou Gehrig, but also such international celebrities as Madame Chiang Kai-shek. The mouse, then, acted as the constant element in two bills that represented fictional China and the documentary one; thus, Mickey was at home on bills presenting all manner of heroes, certainly benefiting from their status but also undoubtedly contributing to it.

Historical Context and the Construction of the Bill

Three film bills with features about World War I or II exemplify the uses of diversity and similarity in addition to some of the interactions of theater experience and cultural context. In August 1938, Paramount reissued its 1932 film, *A Farewell to Arms*. The *Post* reviewer called the movie "rare," referred to the star, Helen Hayes, as "one of the great actresses," claimed that the director, Frank Borzage, "has few equals in directing tender, believable and memorable love stories," and described the supporting cast as "distinguished." Naturally, the reviewer also mentioned the film's literary source and the famous novelist, and said of Ernest Hemingway's book that "nothing more eloquent has been said against war."[34] This, then, constituted a reissue of similar importance to that of *Morocco*; a chance to appreciate once again, in the words of the *Post* reviewer, "a beautiful and an important" story.

Rather than including classics, however, the supplementary bill contained a short called *Unusual Occupations*, "featuring vivid shots of the age-old ceremony of changing of the guard at Buckingham Palace and surprise glimpses of [actor] Reginald Denny as a model-airplane expert." Along with the newsreel there was also "a not-so-good Looney Tune, entitled *Porky's Spring Planting*," and numbers performed by the theater orchestra. The films on the bill were neither generically similar nor thematically alike. There may be a general notion of class common to the feature, based on a great novel and discussing important issues, and the short, which depicted the protection of the British monarchy and the upper-crust English actor Denny relaxing. But the five primary parts of the first-run bill—feature, cartoon, short subject, newsreel, and live act—appear designed primarily to signify entertainment through variety. The antiwar aspect of the feature, therefore, already perhaps blunted by the emphasis on the love story, became simply one of a number of varied elements on the bill rather than the main focus of the narrative of going to the movies.

It may not seem surprising that the Hollywood cinema would seek to deemphasize any overt ideological commentary in an individual film, and that film booking managers would do the same in a group of films playing together. I have stressed the high degree of diversity on the bill, however, because just a year later those in charge of film booking, taking full advantage of the political implications of film content, constructed a bill that seemed specifically aimed at producing an antiwar message.

In December 1939, *Everything Happens at Night*, with Sonja Henie and Ray Milland, opened in Washington. The film sounds like an early version of Hitchcock's *Foreign Correspondent*; as the *Post* review described the plot, "it seems that Dr. Hugo Norden, a Nobel prize winner through his profound advocacy of world peace, was supposed to have been murdered," and as a result, newspaper reporters start "snooping about the Swiss Alps." Finally, however, the very-much-alive Dr. Norden "is spirited aboard a boat bound for America and eternal freedom" from European fascism. Somehow, this setting became the backdrop for a display of Henie's skating, and the reviewer wrote that the movie was merely "a holiday . . . item that will fit snugly into the public's seasonable insistence that it be given nothing to distract attention from the gay implications of a festival period of merrymaking."[35]

Along with a stage show that "accentuates the Christmas spirit," the bill included an extraordinary MGM cartoon, *Peace on Earth*, a film that, unlike the feature, made no effort to disguise its stance toward war. Using typically cute cartoon squirrels and chipmunks, the cartoon explains that humans have become extinct through mindless warfare. Abandoned artillery and helmets serve as the animals' homes, and in one such house an older squirrel tells his grandchildren how humanity destroyed itself. The Christmas spirit suffuses the entire film, so the cartoon has more of a religious sensibility than a political one. Nevertheless, the overt antiwar, apocalyptic message stands out as unusual not simply for a cartoon, but for any Hollywood product.

In just sixteen months, then, something had changed. In late 1939, even a cartoon that was probably put into production before the official outbreak of hostilities in Europe could make a statement about world events, and could be used to bring attention to a well-buried antiwar position in the feature film. If we can assume that the newsreel also mentioned events in Europe, then virtually every aspect of the bill at least alluded to the war, and two of them took positions opposed to it. Therefore, compelling world events, but events in which the United States still did not have a direct involvement, could motivate not only the political content of individual films, but also the content of the theatrical bill. The season itself may have tempered the bill's ideological stance; an examination of theater presentations during holidays might reveal that Christmas provided film companies and theaters with an opportunity, for at least a few weeks, to politicize their offerings precisely because the holiday, by connoting religion rather than ideology, itself seemingly served to depoliticize them.

Once the war was no longer being fought only in Europe, the antiwar bill apparently came to an end, although an overtly prowar theatrical program did not immediately take its place. In June 1942, for example, Harold Lloyd's production of *My Favorite Spy* opened at the RKO-Keith's theater. Starring Kay Kyser and Ellen Drew, the film, according to the *Post*, served up the "combination of the interrupted-bridal-night and trap-the-enemy-agents motifs." This war comedy was no classic, but it played with as diverse a bill as that which accompanied *A Farewell to Arms*: *The Call of the Sea*, "a short subject dedicated to the fishermen and artists' colony of Gloucester," and the Walt Disney cartoon, *Pluto Jr.* Once the United States had entered the war, therefore, the feature itself could be prowar (the honeymoon couple fights the Nazis) but the entire bill would stress variety-as-escape, both literally and figuratively; the live-action short subject removed the spectator to someplace far away, and the cartoon, according to the newspaper review, placed its "emphasis on whimsy." Underscoring the escapist value of the bill, the *Post* concluded that "the newsreel complete[s] a gay and refreshing entertainment wholly on the lighter side."[36] Apparently, then, even the newsreel refrained from dealing with world events. The *Post* sample implies that, at least very early in the United States' participation in the war, the Hollywood product could allude to the fighting, but the theatrical experience sought to deemphasize it.

The three bills with war-film features came from three distinct historical moments; pre–World War II, shortly after the commencement of fighting in Europe, and seven months after the United States' entry into the war. Appearing during a period of intense governmental concern with the war-related content of Hollywood films, they indicate that, if the cinema succeeded in shaping public sentiment about the war, it may not have been because of individual films, but rather as a result of the mode of exhibition itself, that is, the entire theatrical bill.[37]

Although the evidence apparently does not exist—the *Post* stopped providing complete film bill information during the early 1940s—one can guess that, by 1943 or 1944, the social function of the bill became that of promoting much more vigorously absolute support for the war effort, and examining the war itself in much greater detail. Certainly, the content of individual feature films and cartoons (and so, presumably, their combination on the bill) from this period ceaselessly stressed the significance of American war aims.

Indeed, for a large segment of the audience—the male military recruit—this emphasis on "why we fight" became precisely the object of film exhibition. As the following chapter shows, the typical film bill, and the place of the cartoon in it, showed an impressive adaptability in the transfer from domestic to military theaters. Not only did the bill work to legitimate longstanding cultural tensions, but it also attempted the construction of nothing less than a collective identity based on the moral, technological, scientific, and political superiority of the American fighting force.

3 | THE DISAPPEARANCE OF DISSENT

Government Propaganda and the Military Film Bill

More often than not, the first-run film bill in the 1930s expressed all sorts of difference—differences of genre, race, class, gender—as parts of an aesthetic whole, the better to manage those differences, to smooth over the tensions they might cause. Through a particular conception of variety, the film bill turned the theater into a single public place where everything could be seen, a spectatorial utopia seemingly providing a 360-degree view of the world.

Just a few years later, after the United States had entered World War II, another kind of film bill, made for a specific audience, worked to create more overtly a sense of ideological wholeness, and once again to smooth over the potential disturbance of difference. Early in the war effort, the United States government went into propaganda production with the creation of an Information and Education Division. Working with the division, Frank Capra produced the *Why We Fight* series, which, upon completion of the first of the documentaries in 1942, became required viewing for all new recruits. Then, because of their success in asserting United States war aims, Capra's films were released for distribution to a civilian audience.

Far less well known than the *Why We Fight* series is the *Army-Navy Screen Magazine*, produced by Capra for the Army Signal Corps. Beginning in 1943, Capra oversaw a twice-monthly, 20-minute collection of newsreels and special features (usually between three and five per magazine). Although not required viewing, the magazines were difficult to avoid. They were screened at United States military bases around the world, usually accompanying a Hollywood feature film (it is impossible to determine which magazines played with which features, and I would guess that there was no precise system for matching them, with combinations differing from base to base).

Warner Bros. produced twenty-six "Private Snafu" cartoons for Capra's project (fig. 10). Each animated installment ran about three minutes, and usually appeared as the last segment on the *Screen Magazine*. Perceived as harmless because of its association with children's entertainment and whimsical and comic subjects, animation came to be used, in the military theater, as one of the central vehicles of wartime propaganda in order to make that propaganda seem as benign as possible.

The manner in which animation signified a certain kind of entertainment explains not only why it may have been ideally suited to the exigencies of wartime propaganda; it also tells us a great deal about the government's attitude toward its audience. The magazines, and particularly the "Snafu" cartoons with their barracks room humor and pinup girl characters, were directed at a heterosexual male spectator. But despite an audience made up primarily of men (I have not been able to determine whether any women also saw the films), and despite the occasional adult humor, the mode of address was not based simply on gender. Instead, with so much instruction coming from the cartoon segment of the magazines, it seems at least plausible that Capra, along with the other officials at the Signal Corps, believed that the best way to reach recruits was to speak to them as youngsters, as those who might best understand the war through a medium that, while not yet associated solely with children, was certainly believed to be ideally suited to their intellectual level. Indeed, the "Snafu" cartoons' mode of address (many of the shorts employ simple but ingenious rhymes) shifted easily to children's literature during the postwar period. The creator of "Private Snafu," in fact, and one of the series' principal writers, was Theodor Geisel, who had already had a great deal of success writing children's literature and who would become perhaps the most famous author of children's books in the 1950s and 1960s, working under the pen name "Dr. Seuss."

An analysis of the relationships between the short subjects on the magazines, with a special emphasis on the "Snafu" cartoons, explains much about the manner in which entertainment worked to indoctrinate a diverse group of men. Although it is difficult to judge the success of the military's propaganda effort, the magazines themselves tell us interesting things about the military, the media, and the corporation. Influenced by developments in social psychology, the War Department assumed that recruits indeed could be guided by certain kinds of messages. Also, the

Figure 10. Private Snafu, lazy as usual, in *Gas*, from a 1944 *Army-Navy Screen Magazine* (Army Pictorial Service, Signal Corps).

military believed that the type of theatrical bill that had established itself as film entertainment could work just as effectively as a means of government propaganda. Finally, to transmit its messages, the military was able to make use not only of relatively new media technology—enabling films to be made rapidly and exhibited to millions of people—but also of an unprecedented relationship between the government and the film industry.[1]

The "Private Snafu" cartoons are significant for contradictory reasons. First, because they tend to come at the end of the magazines, they often serve to lock in the meaning of any one film bill. That is, the "Snafu" cartoons act as a sort of ideological exclamation point to many of these collections of shorts. On the other hand, however, even as the cartoons provide ideological closure to the magazines, they are also the place where we can read war propaganda against the grain, where we can see the tensions that the magazines tried to manage. The point of most of the cartoons in the "Snafu" series was to show soldiers what would happen when an ordinary recruit did not do his job. But the very insistence on

Snafu's incompetence, which the cartoons equated with his discontent, indicates that discontent must have been widespread, and must also have been one of the primary targets of military propaganda (indeed, the recruit's very name, a soldier's acronym for "situation normal: all fucked up," itself signifies discontent by pointing out the irrationality and disorganization of the military status quo).

The magazines had twin goals, one individual and the other pertaining to the country as a whole: they sought to make any one person's discontent seem aberrant, and to create consensus about U.S. goals during wartime. In working toward these goals, the project of the magazines was nothing less than managing the tensions between individuals and groups, and between various groups themselves, in order to create a national identity during a time of crisis.

Managing Diversity

In 1940, more than 25 percent of all Americans were immigrants or their children, and the diverse ethnic loyalties often determined responses to the possibility of war. As Richard Polenberg has pointed out, German and Italian Americans favored isolation, not out of a fondness for fascism but rather because of an "emotional attachment to their homelands" and a fear of being treated as outcasts if the United States did enter the war. Irish Americans frequently opposed any aid to Great Britain, while Polish, Danish, and Norwegian Americans often supported helping the Allies because their home countries in Europe had been overrun by the Nazis. As just one more example among many, Anglo-Saxon Protestants, with strong ancestral links to England, were among the most vocal interventionists.[2]

Opposition to the war had as much to do with race and class as with ethnicity. In a poll taken in 1942, 18 percent of Harlem's African-American population felt that they would receive better treatment if Japan won the war, and 31 percent felt that their treatment would remain the same whether Japan or the United States was victorious (only 28 percent felt that a Japanese victory would make things worse).[3] Before Pearl Harbor, many labor leaders, with the UMW's John L. Lewis most prominent among them, insisted that the United States' entry into the war would pit an American working class against a German one in a fight that would result in propping up one aristocracy at the expense of another. Of course, for working-class soldiers collective resistance within the military and concern with labor problems at home probably became casualties of the severity of military discipline and the typical recruit's overriding desire for self-preservation. And military propaganda itself worked to make those tensions not necessarily disappear, but undergo a sea-change of signification. Issues of labor versus capital, for instance, were transformed into issues of the selfish slacker versus the group. Class tensions frequently were highlighted, not to criticize them but to naturalize them, to make hierarchies a military necessity. Gender differences became a means of showing the relationship between the homefront and the theaters of war. And, of course, all of these differences were made to seem minor in comparison with the conflict between the Allies and the Axis.

In the 1930s, the first-run film bill functioned to create a space where everything could be seen. In 1940s military propaganda, the film bill constructed a space where much might be displayed, but only one thing, finally, was on view: a national consensus accepting hardship and responsibility in order to achieve an American-made peace. To control this view, the *Screen Magazine* appealed to two forms of power: that of medicine and science, and that of a controlling male gaze. Both together constituted systems of rank, hierarchy, collective good, and discipline, and also informed audiences how to think about class, gender, religion, technology, leisure, and home.

While rarely thematic—that is, covering a single issue—the magazines used the various combinations of newsreel, special segment, and cartoon to fix attention on one of several core issues. Some of the magazines constructed a scientific discourse that assured audiences of both the benevolence of the United States government and the superiority of its technology. Others insisted that biology determined behavior, making the military hierarchy seem natural and therefore beyond questioning. Still other magazines emphasized the hardships on the homefront and the necessity of fighting the war there as well as in Europe and the Pacific, and implied that a soldier's discontent meant that he was letting down not only his comrades in arms, but also the civilians back in the United States. In related fashion, some magazines, rather than militarizing the homefront, sought to make all military activities seem related to the war effort so that even a soldier's leisure must be directed toward defeating the Germans and the Japanese.

Thus, the *Magazine* stressed the role of the individual within the group and asserted unity within a strictly hierarchical system. Insisting on individual responsibility—despite the superiority of American technology, for instance, it was the task of each soldier to use his equipment and weaponry properly—the magazines insisted also that discontent itself was not only individualized and therefore not characteristic of the enlisted troops, but also abnormal. The *Magazine* sought the creation of a consensus military culture, a culture in which difference must be dangerous and unity the greater good.

Militarizing Culture

As an instrument for the dissemination of government policy, the magazines marked a beginning stage in what sociologist Harold Lasswell, in the title of his influential 1946 article, described as "The Garrison State." For Lasswell, in such a state distinctions between the military and civilian life would be blurred, and all individual and social efforts would be directed toward effective militarization of all aspects of culture. According to Charles Moskos, "the garrison state would be characterized by the militarization of the civil order as the military system became coterminous with the larger society."[4] Both World War II and improved media technology presented the federal government and the War Department with their first opportunity to address, over a number of years, millions of people who were employed by the military but who identified primarily with their previous civilian status. Like the *Why We Fight* series, the *Army-Navy Screen Magazine* served to make the transition from civilian to soldier, and then back again, seem part of a larger system of unquestioned allegiance and eternal vigilance.

In one sense, the magazines were apparently value-neutral. Capra himself encouraged the notion that the magazines, rather than embodying official policy, actually sprang from the soldiers themselves. Through its "By Request" feature the magazine encouraged troops to send in ideas for segments. John Laffin, in his history of American troops, writes that "one soldier wanted to hear a quartet sing 'Down by the Old Mill Stream,'" and that Capra and his crew "met this and a thousand other . . . requests."[5] *Documentary Film Classics*, a catalogue produced by the National Archives, adds that "troops most frequently asked for scenes of their home towns or families."[6]

Despite this appearance of directly expressing soldiers' desires, the magazines in general and the "Private Snafu" cartoons in particular were, in the manner of so many Hollywood films from the period, "avowedly discursive," to use Dana Polan's term. Each cartoon announced itself "as a message from somewhere . . . to someone."[7] A credit reading "Official Film, War Department," opens each of the cartoons, and then a closeup of Snafu moves toward the audience, with the star's name under the photo (fig. 11). These opening two shots established a link between the government and the star, leaving no doubt as to the source of the film; the message came from the government through the star, who himself signified all recruits. In this case, then, and despite the strategy of the "By Request" segments, the "Private Snafu" cartoons clearly came from the government, and reached the enlisted-man audience through the intermediary of the star.

DEMOGRAPHICS

The armed forces during World War II provide us with relatively precise information about audience. We can, then, make a fairly educated guess about who viewed the *Army-Navy Screen Magazine*, and which spectators the government targeted for its propaganda. Indeed, the War Department's fascination with the possibility of social engineering in the armed forces, along with the development of more sophisticated demographic methods, led to a remarkable recording of who served in the military during the war.

The audience itself was immense. In 1939, the military comprised only one percent of the country's male labor force. By 1945, however, it counted for 20 percent of the United States' *total* labor force.[8] In the four years of United States involvement in the war, about 14 million men served in the various branches of the military. While I have not been able to locate the exact number of those serving between 1943 and 1945, the years of *Screen Magazine* production, the military did record some detailed information about the men in the service during that period.

In 1943, 42 percent of all enlisted men were twenty to twenty-four years old, 54 percent had a high school or college education, 90 percent

Figure 11. How Private Snafu appeared in the credits of the cartoons in the *Army-Navy Screen Magazine* (Army Pictorial Service, Signal Corps).

were white, and 8 percent black. Sixty-one percent of the men came from the North, 29 percent from the South, and 10 percent from the West. Throughout the war period, approximately 5 percent of enlistees were Jewish.[9]

Black Americans entered the military in large numbers during the war, but the 1943 statistic cited above remained fairly constant, while an army study from the same year revealed that 84 percent of white soldiers opposed military integration.[10] Segregation was still the rule, and units were frequently organized around race; an Eskimo unit, for instance, saw action in the Aleutians.[11] In the army, almost 40 percent of all troops were trained for ground combat, while about 20 percent were service workers and laborers, 15 percent worked in administrative capacities and mechanical or craft jobs, 10 percent in technical and scientific positions, and another 15 percent in a variety of other capacities. In terms of economic class, 70 percent of enlisted men occupied the three lowest pay grade levels: E-1, E-2, and E-3.[12] Indeed, the military class hierarchy would become an extremely important issue within the armed forces, and therefore an extremely important issue in military propaganda.

In *The American Enlisted Man*, Charles Moskos writes that although it has become an aspect of military mythology that "total regimentation precludes internal conflict," this conflict always exists within American armed forces organizations. Moskos records intense antagonisms not only between enlisted men and officers, but between lower-ranking enlisted men and noncommissioned officers. The United States' entrance into World War II and the concomitant massive conscription program only exacerbated the problem by making temporary soldiers out of millions of civilians who often came from a higher class—whether economic or social or both—and had received more education than many of the career NCOs who trained them. As Moskos writes, "the enlisted culture is derived from a social organization which under-utilizes . . . middle-class individuals while simultaneously allowing persons from lower- or working-class background to participate with minimal acknowledgement of preexisting socioeducational handicaps."[13] I do not mean to criticize this democratization of power; it is important to point out, however, that it brings with it class tensions and antagonisms fairly specific to the military experience.

Along with the problems brought on by class issues, the military also had to deal with a large number of soldiers, and perhaps a majority, who had no desire to be fighting in the first place. Most of the troops had been drafted or had volunteered because they were about to be drafted anyway. As a result, the troops were motivated not by patriotic goals, but by pragmatism.[14] Thus, the *Army-Navy Screen Magazine* addressed an audience that was overwhelmingly male, primarily Christian, predominantly white, and significantly disgruntled.

The films on the magazine bills reflected all of these characteristics. They concentrated on male wartime activity, except for those newsreels that dealt with changes on the homefront, most notably the seeming feminization of the workforce, or those features that provided cheesecake ("Phil Spitalny and his all-girl orchestra play selections from *Oklahoma*" in Number 22, from 1944, for example). Christmas became the holiday-of-choice in those magazines that dealt with special occasions. The celebrity/heroes almost always were white; movie star Joe E. Brown, for instance, or baseball players Red Ruffing and Joe DiMaggio in Number 13, from 1943 (and, indeed, Snafu himself, the representative of all enlisted men, was white). Finally, and most significant to the issues that concern me here, the films in the magazines in general, and the "Private Snafu" cartoons in particular, worked to assuage discontent.

The Corporation and the State

Made by Warner Bros., the *Army-Navy Screen Magazine* demonstrates the symbiotic relationship, formed during the war and continued afterward, between corporations and the government. As George Lipsitz has pointed out, an executive elite emerged as the true victors of the war, an elite created by wartime government spending that went to a "relatively small number of companies [that] derived most of the benefits from federal expenditures during the war." According to Lipsitz, two-thirds of government business expenditures (about 117 billion dollars) went to only 100 companies, with the rest being spread out over an additional 18,000 companies. Those most-favored corporations benefited from huge tax breaks besides.[15]

I do not mean to imply here that Warner Bros. stood out as one of those privileged companies. The studio's production of the "Snafu" cartoons, however, does indicate the extent to which government practice and corporate practice merged during the war. Demonstrating the link between patriotism and the profit motive, all of the major studios were happy to do contract work for the government, particularly during a period of relative financial uncertainty (some European markets, for instance, had been closed to Hollywood films). The studios gave over a great deal of their nonfeature film production to government purposes—training films, war bond shorts, and, in the case of Warner Bros. and Private Snafu, cartoons. Further complicating the issue, at least in this one case of cooperation, the government and the movie studio hardly worked as equal partners. Studio assistance probably came about as much from political necessity as patriotic fervor or a desire to increase profits; quite simply, the studios of this era often felt compelled to work with the government in order to prevent, or at least forestall, government investigation of major studio monopoly practice.

Both in terms of production practice and subject matter, then, the films on the *Army-Navy Screen Magazine* bill concerned themselves with power. The various kinds of discipline represented in the films were turned back on their audience, as the recruits learned just how the government expected them to act, and, indeed, how they must act, for the proper prosecution of the war. Moreover, both implicitly and explicitly, the magazines demonstrated the power of the government, which announced its role as producer at the beginning of the "Snafu" films, to regulate behavior in the military, and the power of the corporation, unnamed in the magazines, to produce the representations of governmental desire.

The Science of Persuasion

The militarization of American culture, and the blurring of distinctions between commerce and government, took place within a context of new interest in the possibilities of propaganda. Jacques Ellul has explained how, during the 1920s and early 1930s, the main emphasis in propaganda study was on the subconscious, on propaganda "as a manipulation of psychological symbols having goals of which the *listener is unaware*" (emphasis in original).[16] Several years later, however, psychologists came to believe that the most effective propaganda was the most overt; propaganda should state its objectives clearly. This shift corresponded to a new wartime enthusiasm on the part of the government and the military for social engineering, an increasingly popular and credible branch of psychology that had grown out of behaviorial psychology, and that argued that "human behaviour can be manipulated towards socially desirable goals."[17]

This confidence in social engineering along with advances in media technology encouraged the War Department to produce the *Why We Fight* films and the *Army-Navy Screen Magazine*. But besides being enthusiastic about these new methods and technologies, the government was probably concerned with combating propaganda from other sources. In his 1940 psychological study of mass persuasion, *Conquering the Man in the Street*, Ellis Freeman wrote, "when a large number of persons combine for a common purpose, each one feels less aware of his own weakness and limitations," and that "they can together achieve much more than the mere sum of their accomplishments separately." Freeman added that when these persons "are united under a leader, he will come to personify them collectively, to embody their aspirations, and to get most of the credit for accomplishments." Referring primarily to the growth of fascism in Europe, the author stressed that Americans, because of their skepticism, possessed a much greater "sales-resistance to tendentious indoctrination."[18] Nevertheless, Freeman cautioned that despite this resistance, "clever propaganda [could] get around it from another side after having lulled us into a false sense of security."[19] From the rather apocalyptic title of Freeman's book to its emphasis on fascism, we can get a sense that, during this period, a generalized fear existed in the United States of Italian and German fascist ideology imperceptibly winning over American citizens. In fact, Freeman referred with some frequency to Huey Long and Father Charles Coughlin as cautionary examples of the possibilities for fascism at home.

Chapter 5 will explain how the federal government, through Nelson Rockefeller's Office of Inter-American Affairs, went into propaganda production and dissemination in South America during World War II in order to combat Nazi propaganda in the area. It is not surprising, then, that the War Department, almost certainly aware of studies like Freeman's, decided to use some of the tools described in these studies on a homegrown audience—in this case, those millions of men in the armed services—in order to convince them of the correctness of fighting fascism. At least according to official pronouncements, the government's efforts were successful. *Documentary Film Classics*, the previously mentioned catalogue from the National Archives, claims that "by the end of the war, the *Army-Navy Screen Magazine* reached an enormous weekly audience of 4.2 million," and that it had become "for American soldiers all over the world a communal experience that greatly influenced their perception of the war." The archive catalogue concludes that "although the *Screen Magazine* was, of course, an instrument of official policy, its honest no-nonsense approach won it the respect and affection of its GI audience."[20]

In part because of the National Archives' preservation policy, we can analyze this no-nonsense approach through the construction of the military film bill and the place of the cartoon on it. All of the magazines are available at the Archives, and all of them are indexed and described in card catalogues; throughout this chapter, the quoted descriptions of the magazines will come from these cards.

Military and Scientific Authority

Many issues of the *Magazine* deal directly with the role of authority during wartime, stressing government benevolence in general, and, more particularly, the necessity of the military chain of command. The final segment in Number 24, for instance, from 1944, provides a convenient narrative closure to the *Magazine*, and also a fantasy closure to the war itself: newsreel footage shows Japanese warships, in action near the Marshall Islands, being strafed from the air and Japanese planes crashing into the ocean, followed by a Japanese surrender. As the index card at the National Archives describes it, the Japanese prisoners "receive food and medical care." Here, the end of the *Magazine* posits the United States as a fighting force nonpareil, but also as a benevolent conqueror giving aid to a defeated enemy.

Immediately before the newsreel, however, the audience watched a cartoon concerning "the contracting of diarrhea by Pvt. Snafu, who refused to wash his mess gear." Hardly an unusual predicament for Snafu, this one serves to point out one of the interesting paradoxes in the magazines' construction of prowar propaganda. Underscoring the high moral values that American troops defended against Japanese and German aggression, the *Magazine* depicted the United States as globally benevolent, protecting even its enemies. On the other hand, the individual Americans drafted in order to assure their government's mission were individually responsible for their own well-being. The War Department would provide them with state-of-the-art fighting equipment, but it was up to the soldier to make the proper use of it. The *Magazine* thus constructed relations between micro and macro levels. Had Snafu kept his gear clean, he would have remained healthy. Had he remained healthy, he could have fought for the moral order his country represented. Actual war objectives became obscured by the moral superiority of the Allies, which, in turn, were dependent on the behavior of even the lowest private. Rather than simply stressing pulling together, then, government-produced war propaganda insisted that even those who were powerless—draftees like Snafu—possessed power.

This formulation of a medical discourse bound up with the U.S. government on the one hand and individual responsibility on the other also figures in other magazines. In Number 23 (1944), right after wounded American fighter pilots are removed from their planes and given medical aid, Private Snafu, in *Private Snafu versus Malaria Mike*, dies after ignoring antimalaria measures. At the end of the cartoon, as a trophy mounted on the wall of his mosquito nemesis, the late Snafu speaks to the soldier/spectator: "Our program has come to you through the courtesy of my sponsors, the United States army, distributors of GI repellent, mosquito net, xxxxx tablets, and good old fashioned horse sense. Gee, I wish to hell

I'd used them." Once again, films on the *Magazine* bill contrasted a watchful, providing government with the short-sighted, irresponsible individual.

In Number 52 (1945), Anopheles Annie, the mosquito-star of *It's Murder, She Says*, "brags of her conquests and laments anti-malaria measures, but finds Private Snafu to be a good prospective victim." The same issue also shows "the talent audition opportunity which NBC [Radio] offers returning veterans," followed by soldiers liberated from a Japanese prison camp recounting the Bataan death march. Number 38, from 1944, begins with *Target Snafu*, which "shows by animation a malaria mosquito squadron attacking Private Snafu who is hospitalized in a malaria ward." Then soldiers saw a newsreel that took up the rest of the *Magazine*, and showed Allied tanks liberating Paris "as civilians cheer."

These "Snafu" cartoons indicate the threat posed by malaria to American forces. But within the context of the *Magazine* series, they explain something about World War II propaganda, and the place of science and industry within it. Dana Polan has written that "so much forties science distinguishes itself not by modesty or entrenchment—a retreat into specialization—but rather by vigorous advance, a panoptic desire to gaze upon everything, to see all events as falling under its purview."[21] We see precisely that panoptic gaze in these magazines. In them, medicine has become a testing ground for United States superiority, both scientifically and morally; the government provides amply for its own soldiers, and cures a Japanese enemy that, in Bataan, marched its prisoners until most of them died. But this superior science fails when even the most minor participants in the war—Private Snafu, for instance—ignore their own responsibility in the practice of that science.

Medical discourse in the magazines helps to establish the moral backbone of the United States along with the moral responsibility of the individual soldier. Moreover, it equates science with military endeavor itself: Snafu, in that hospital bed, must be defended against a squadron of mosquitos, after which military might shifts from the sickroom to the battleground as those tanks liberate Paris. Medical discourse finally means industrial discourse, as the preventive treatment of potential malaria victims in the "Snafu" cartoon finds itself paired with NBC Radio's treatment of returning veterans. The soldier/spectator, then, could equate one kind of cure with the other, with medical science helping him survive the war and industrial science helping him adjust to peacetime.

The films in the magazines show diverse aspects of a new world, one born out of war but seeking peace, and regulated by an all-encompassing science. Numerous issues of the *Magazine* insist on the possibility of better living through U.S. science and technology, and in doing so reverse the warning of so much popular culture from the 1930s. During that decade, danger frequently came from mad doctors trying to push science into God's realm (Dr. Frankenstein, for instance, in the 1931 film version of Mary Shelley's novel). Just a few years later, however, the example of such lesser figures as the negligent Snafu posited that disaster came only when normal citizens disregarded science as the solution to all problems.

In their attempt to construct a wartime consensus, the magazines implied a democratized responsibility, beginning with privates and going up the chain of command. The combination of cartoons and newsreel footage also suggested that despite their individual tasks, everyone in wartime was bound by the highest of all moral purposes—Western religion. The second segment of Number 29 (1944), for example, shows that "troops struggling through a South Pacific jungle pause for chaplain services and then continue." In *The Chow Hound*, however, Snafu "wastes food by taking too much." Although perhaps accidental, the contrast nevertheless must have been obvious. Beleagured soldiers assert their collectivity, and the cause for which they fight, through their Christianity, while Snafu, safely behind the battle lines, signifies only selfish individuality by overstocking on rations.

Class, Gender, and Biology

Although prioritizing the collective, the magazines carefully insisted upon a military social hierarchy that was biologically determined. Precise demographic information is difficult to come by, but we can assume that the enlisted men watching the magazines were themselves often lower middle and lower class. The enlisted everyman himself, Private Snafu, was clearly marked as lower class through his stereotypical Brooklyn accent and poor grammar. Regardless of demographic reality, therefore, the *Magazine* represented the recruit as poorly spoken, poorly educated, and generally unintelligent. As a result, in this instance, the magazines functioned slightly differently from the first-run bills in Washington during the 1930s. Those showed different class experiences in order to commodify class difference and so control the problems caused by class distinctions. The magazines also stressed class diversity, but in order to make the distinctions seem necessary to the proper execution of the war.

For example, Number 5 (1943) shows newsreel footage of WACs being reviewed by General George C. Marshall and Secretary of War Henry L. Stimson. Following the newsreel, Snafu, in *Gripes*, complains about his kitchen duty and has his wish for upward military mobility granted by a Technical Fairy First Class. Of course, Snafu in charge of the base equals disaster; as the National Archives index card explains it, Snafu "permits discipline to break down; the enemy attacks unopposed."

The newsreel in Number 5 uses the typical cinematic construction of the gaze—men looking at women—and aligns it with unquestioned authority. Two male representatives of military authority, one a member of the army and the other a civilian, are allowed to watch whole squadrons of women. What usually functions in the cinema as a privilege of gender (women in movies far less often get to gaze unproblematically at men) becomes here a privilege of class; only those men high up in the military command get to review women. The newsreel implies that this is as it should be, because reviews are standard military procedure and generals and secretaries of war are the natural people to engage in it.[22]

The cartoon accepts and constructs this naturalness by showing what happens when someone who should not have the privilege of command actually gets it. And, perhaps in large part because the audience for the magazines was a male one, the threat to the military hierarchy does not come from women, but from the men themselves. More specifically, the threat comes from a lower-class man aspiring to rights enjoyed by those above him. Significantly, too, the *Magazine* implies that this military hierarchy, which one could assume to have less to do with birth and more to do with achievement and ability, did indeed match the class hierarchy in the civilian world. Possessing many of the stereotypical attributes of the lower class, Private Snafu occupies the lowest rung on the military ladder, while the distinguished Cabinet member Stimson (who himself represents the absent president, the aristocratic Franklin Roosevelt), has attained the highest.

In different form, *Magazine* audiences received the same messages as civilians. Domestically, the war period was marked by the determined efforts of the Roosevelt administration to convince workers to lessen their demands voluntarily for the duration. Responding to prewar protests by many labor leaders that entry into the European conflict would simply mean an American working class fighting a European one for the benefit of an entrenched but embattled monied class, Roosevelt stressed the necessity, during the war, for all domestic workers to give up individual interests for the common good. In other words, and probably no differently from most modern governments during times of war, the administration attempted to convince labor that what was good for the world necessarily was good for labor.

The magazines assert a similar national discourse of patriotic duty. Before the war, many in the audience would have been much more aligned with labor than with capital. As a result, a "Snafu" cartoon such as *Gripes* tried to convince this part of the audience that their discontent was at best merely selfish (the title certainly underscores this) and potentially even dangerous. Moreover, the overall *Screen Magazine* bill demonstrated the correctness of a structure that stressed absolute differences between classes, and which made kitchen duty as much the proper preserve of privates as reviewing the troops was for generals and Cabinet members. The Roosevelt administration, then, tried to teach civilian labor not to strike, and enlisted labor not to gripe.

The relationship between propaganda production and domestic policy is not simply that of a direct cause-and-effect link. Yet, it is perhaps not insignificant that the dozen years or so between the Second New Deal of 1935–1936 and the immediate postwar period witnessed an increasingly reactionary government stance toward labor. The *Army-Navy Screen Magazine*, along with the Roosevelt administration's position on labor unrest, made for a stark contrast from the progressive, prolabor, and Roosevelt-supported Wagner Act of 1935, and presaged the passage of the sternly anti-union Taft-Hartley Act of 1947. Similar to the representation of science during the war, what began, apparently, out of wartime necessity served to initiate postwar domestic ideology.[23]

It would overstate the case to insist that the various editions of the *Army-Navy Screen Magazine* addressed themselves to a typical working man who could be influenced to accept class separations in the military because they matched precisely the divisions in civilian life. As mentioned earlier, during World War II and for the first time in United States history, many of the men at the lowest end of the armed service hierarchy—the privates—arguably came from a higher class than the men directly above them—the sergeants—who instructed them in military protocol.

Perhaps as a result of this confused hierarchy, the magazines tend to deal only with generals and privates, from Marshall to Snafu (in contrast, many of the wartime cartoons made for a domestic audience, such as

Donald Gets Drafted [Disney, 1942] or *Ace in the Hole* [Universal, 1942] emphasize the brutish drill sergeant). By themselves, the "Snafu" cartoons ignored military demographic reality, and implied that all privates were poorly spoken and semiliterate. But even if the average enlisted man viewed himself as superior to the doggedly incompetent Snafu, the magazines stressed an absolute subservience to authority and a biological determinism to military positions. Privates as well as generals were born and not made.

Gripes, for example, implies that upward movement, at least for privates, goes against a military law of nature. Other issues of the *Magazine*, through the combination of films, encouraged the same belief and asserted that even lateral movement subverts the natural order of things. Number 11, for instance, from 1943, begins with a newsreel of Admiral William Halsey overseeing the landing of a naval convoy and ends with Snafu in *Infantry Blues*. In another appearance in the series, Technical Fairy First Class grants Snafu's wish to leave the infantry and transfers him to a tank corps and then to the navy. Of course, the results of this mobility are disastrous, and Snafu learns his lesson: that all branches of the military are difficult, and he should stay in that branch to which he is best suited. This cartoon at the end of the *Magazine*, therefore, locks in one of the possible readings for the opening newsreel. It was Admiral Halsey's birthright to lead that convoy, because the cartoon demonstrated that the military assigned position according to natural ability and inclination. Like other magazines, this one underscored the illogic of challenging military authority.

By insisting on the military job-as-fulfillment-of-destiny, the magazines made use of a widespread wartime discursive strategy that emphasized biology as an explanation for action. Lending scientific support, one psychological study of soldiers who went AWOL asserted that 63 percent of them suffered from mental deficiencies; another study of one hundred AWOL soldiers determined that "twenty-five were mentally defective, [and] the average IQ for all one hundred was 81.5."[24] The magazines could also use an implied biological model to demonstrate the irregularity of women in the wartime workforce. Between the newsreel of Halsey and the "Snafu" cartoon, the *Magazine* showed how, at home, "women pick fruit and drive tractors on a farm," and then how "children assist in milking and churning." By aligning women with children as wartime workers, the *Magazine* indicated that women had no real place in the workforce, but, like children, must work for the duration out of patriotic duty. Thus, the *Magazine* told its audience (who, indeed, had been subject to a kind of natural selection, as a significant number of all draftees was refused induction for various medical and personal reasons)[25] that the government had merely asked them to do what they were best at in order to bring the war to a close. Then, upon their return to the United States, these recruits would help to reassert traditional employment practices at home—that is, employment practices that excluded women from typically male jobs.

The Homefront

The magazines frequently demonstrated the necessity, explored in Number 5, of turning women and children into temporary workers, and stressed the importance of fighting the war not only in the Pacific and European theaters, but at home. Virtually all of Number 31, from 1944, juxtaposed images of home and away. The first newsreel showed Marlene Dietrich and Irving Berlin at a Fifth Army radio station, and then the men in the field who listened to them on portable radios. The next film in the *Magazine* documented the "problems confronting the veteran returning to civilian life," and then explained "mustering-out pay and job training available to veterans." At the end of the *Magazine*, the "Snafu" cartoon, *Censored*, depicted the potential dangers of violating army censorship rules. Writing to his girlfriend back home, Snafu describes troop movements. In what turns out to be Snafu's nightmare, the information falls into Japanese hands, allowing the enemy to defeat the United States forces on Bingo-Bango Island.

From this, we can see one of the major representational strategies of government-sponsored propaganda from the latter part of the war. The homefront must be depicted not only as something worth fighting for, but also as a place worth returning to. Largely through technology, in this case the portable radio, the soldier can carry a little piece of home with him—Marlene Dietrich and Irving Berlin (who themselves show the complexity of wartime signification: the first standing for [naturalized] American glamour, the other for homely values). Further, it is the government itself, through pay and job training, that allowed the veteran to adjust to life back at home. Conscription and the war itself, which took millions of men away from home, turn into the means through which men become settled and comfortable back home. The cartoon, however, proved that, even as the war wound down, vigilance and subservience to military authority were still necessary, and this applied not only to European, Pacific, and African battlefields but also to the domestic front. There were spies everywhere, making the home not a haven from the war but an extension of it. The homefront, according to the *Magazine* in a precursor of 1950s-style anti-Communist doctrine, was a place where veterans could get back to their real lives, but where those lives always would be informed by the exigencies of war.

Closer to the end of the war, Number 42, from 1944, reinforced the war-at-home motif. In the opening newsreel, "J. Edgar Hoover describes espionage activities and security measures taken against [domestic enemies]." Then, audiences saw "bundist camps in the U.S.," followed by "dramatized" sabotage activities and a report on "the presence of German submarines off the Atlantic coast." Prior to the cartoon, every part of the *Magazine* bill emphasized the shift in the battlefield from over there to the homefront. In *Three Brothers*, at the conclusion of the *Magazine*, Snafu must sort shoes, and begins envying his civilian brothers "for their soft jobs." Technical Fairy First Class appears once again and shows Snafu precisely what those brothers are doing: one helps to train war dogs and the other is "facing many trials as a farmer."

When the cartoon ends, Snafu has a new respect for the problems of those men who stayed home. Very probably, the audience also reached an understanding from the entire *Magazine*, not only of the everyday difficulties faced by men like Snafu's brothers, but of the constant danger from a hidden, foreign enemy, and from domestic spies. In Number 42, traditional discourse about the war becomes transposed in full to the homefront, as Snafu's job as soldier—sorting shoes—seems even less militarized than the implied job of all citizens at home to defend against the bundists and the Germans.

Keeping the homefront safe and strong becomes conflated in the magazines with the possibility of possessing a home. Number 36, for example, also from 1944, shows "the farm labor shortage." It depicts an entire farm family pulling together in order to keep their business functioning, as the farmer's wife "tends the poultry" and "his daughter drives a tractor." In *Payday*, the cartoon that follows, Snafu wastes all of his money on trinkets and women, after which Technical Fairy First Class informs him that he has lost his chance of ever buying a house. In this juxtaposition of films, the *Magazine* posits the war as deadly serious (at home) versus the war as potential spending spree (overseas).

Just as with Snafu's brother in Number 42, this issue of the *Magazine* uses the farmer as the representative of domestic issues. The home has become linked to traditional agrarian, Jeffersonian values, and the *Magazine* implies that these values must be maintained. The greatest threat, however, comes from wasting one's chance for upward mobility and economic security, signified by home ownership. Having mythologized and made heroic the agrarian values, the *Magazine* then insists that the war must be fought to defend one's right to become a member of an expanding urban and suburban middle class. According to the *Magazine*, it is the war itself and the employment offered by the government that allows citizens like Snafu even to contemplate middle-classness. As a result of the link between consumerism and war, domestic comfort and home ownership become the fulfillment of a soldier's patriotic duty. Then, in a reprise of the medical discourse of other magazines, individual actions (Snafu throws away his money) rather than social forces preempt the possibility for class mobility.

Militarizing Leisure

Snafu's never purchased but now lost home in the suburbs signified not only class ascendance, but also the unrestricted leisure time that was coming to be mythologized as part of middle-class existence. In dealing with representations of leisure, the magazines began the process of working out one of the contradictions of postwar ideology: the apparent contrast between the constant need for vigilance at home and the increased leisure and comfort brought about by the demise of the twin evils of the 1930s, fascism and economic depression.

Magazine Number 9, from 1943, for example, presents newsreel footage of "Russian pilots [listening] to a pianist." Then, "several pilots take off while the concert continues," and "two U.S. soldiers meet an Australian officer at the Empire State Bldg.," where "they discuss the Bismarck Sea Battle." The "Snafu" cartoon comes last, after scenes of "women working in a machine shop." In *The Goldbrick*, Snafu tries to avoid doing drills; he prefers staying in bed. But as a result of not practicing his drills he is in poor physical condition when his platoon falls under Japanese attack.

The newsreel in the magazine clearly aligns tourism and cultural pursuits with defense activities. Enjoying leisure while on the job, the pilots, in a scene that conjures the opening images of German aircraft in *Triumph of the Will*, mix bombs and Brahms as it were, while the soldiers on leave and taking in the sights nevertheless discuss recent military history. In contrast, Snafu would rather sleep than fight. Because this leisure would serve to exempt Snafu from his work as a soldier, the cartoon makes Snafu seem inferior to the women who willingly worked as machinists. And in contrast to the pilots and soldiers, Snafu engages in an unproductive leisure activity—he is, simply, lazy—one that has nothing to do with furthering the war effort.

Similarly, Number 19 (1944) begins with United States troops debarking in Iceland, and then shows "planes patroling wreckage of a German plane, quonset huts, spring mud, a recreation center, and soldiers on sentry duty and skiing." Leisure and work have become blurred as skiing and patroling both become significant parts of a day's activities, and the troops spend their time divided between quonset huts and a recreation center. A benevolent army has provided the opportunity for ample leisure activity, but only in the context of winning the war and with the emphasis on collective identity rather than selfish indulgence.

Another newsreel in the *Magazine* shows highlights of the Army-Navy football game, so that, for the audience of enlisted men, even a typical masculine activity—watching football—became charged with military significance: the rivalry between the service branches and the physical toughness of the future officer corps. After the game highlights, audiences watched Snafu in *Booby Traps*, in which, encountering a harem while wandering in the desert, the star serenades the seraglio with "Those Endearing Young Charms," accompanying himself on the piano. Of course, because Snafu has disregarded his booby-trap training, the piano explodes when he hits a specially rigged key. Snafu's highly sexualized and completely demilitarized leisure has resulted in a triumph for the enemy, who, as in so many films in the *Magazine* series, lurks everywhere. The model for the soldier-spectator, therefore, was the good leisure of the beginning, both the leisure-as-vigilance (skiing while on duty), and the leisure that is always informed by its connection to the military (the Army-Navy football game).

Technology and Business

Just as important in the magazines as individual vigilance was the system of national production and technological innovation that a modernized America produced during the war. In Number 9, those women working in the machine shop, and also the footage of "a propeller in production

[and] a ship launching," make a stark contrast to Snafu in *The Goldbrick*, in which the star, refusing to work, brings on a Japanese victory. Then, in Number 15 (1943), the newsreel at the beginning shows locomotives being repaired and operated and also the repair of bombed train tracks. At the end, Snafu, stationed in the Arctic in *The Home Front*, "learns by magic television that U.S. citizens are working wholeheartedly for the war effort." Here, on a nuts and bolts level, U.S. technology functions therapeutically, fixing those trains and tracks. But it also works magically and in unimagined ways, as television technology brings comfort to the faraway soldier.

While celebrating the modernity of American technology, this depiction also stresses its fundamentally benign nature. That is, it brings the world to Snafu, helping remedy his homesickness. This wartime insistence on therapeutic and benevolent technology and industry had significance for more than just the lonely soldier at some farflung outpost. The war period witnessed relatively unrestrained industrial growth because of the cooperation between the government and corporate management. George Lipsitz has documented the ways in which government and business worked together to decrease worker unrest in order to stimulate wartime production, and also how a corporate elite emerged from the war to take a previously unheard-of control over the government and the economy. As just one example among many, Lipsitz cites how, by the end of the war, "America's 250 largest corporations operated 79 percent of all new, privately-operated plant facilities built with federal funds," and that "the facility holdings of these few corporations by 1945 equalled the facility holdings of *all* corporations in 1939" (emphasis in original).[26]

The *Army-Navy Screen Magazine* deemphasized not only the wartime profits achieved by business, but also the political power that those businesses had won. Implicitly, the magazines let audiences know that the United States' entry into World War II, and the concomitant increases in production, had ended the Depression and had created ample jobs. Lost, in the midst of the magazines' overdetermination of the marvel of industry, were the limits on wages faced by wartime workers, or the unsafe working conditions, or the collusion between unions, management, and government, all in what Lipsitz calls "the interest of uninterrupted production."[27] Among those wholehearted workers in the *Home Front* are Snafu's father, building tanks, and his grandfather, riveting battleships, the animated equivalents to the women in the machine shop in Number 5 and the workmen who repair the locomotive earlier in Number 15.[28]

The Narrative of Discontent

Issues of the *Magazine* operated like so many wartime narratives, and particularly film narratives. They could acknowledge conflict—the omnipresence of the enemy, for instance, or Snafu's resistance to the military hierarchy. Ultimately, however, like so many popular narratives from the war period, they had to provide a comfortable closure to that tension, a resolution that truly resolved things, that settled all problems, and usually in the name of a greater, patriotic good.[29] That resolution, for the magazines and for so many other films (and also undoubtedly for the popular culture of most of the countries fighting the war) was a very literal one—the necessity for everyone to pull together for the ultimate closure, the end of the war.

Of no small interest is that, within the context of *Magazine* bills that so affirm that unity, the "Snafu" cartoons provide a representation of discontent, of resistance. In practically all of the cartoons Snafu has problems with military discipline or military technology. Of course, the cartoons always view this problem as aberrant, but the insistence on eliminating it indicates that it must have been more widespread, more normal than the government would have liked to acknowledge. Thus the "Snafu" cartoons and the magazines in which they appeared worked much like the film bills from the *Washington Post* sample. Those bills contained class and ethnic difference by commodifying them. More than one class or race on a bill came to signify "variety" rather than act as a representation of a class-and-race-stratified society. Similarly, the "Snafu" cartoons, by themselves, made the star's resistance abnormal. The issues of the *Magazine* in which the cartoons were embedded, and which stressed heroism, production, and authority, worked to naturalize an ideology of unity, and to make Snafu's behavior seem at best unintelligent, and at worst un-American.

My reading of the *Army-Navy Screen Magazine* reveals some of the structuring discourses of government-produced wartime propaganda. A perceived need to create consensus made the government, through War Department film production, attempt the construction of a national identity, a shared sense of what it meant to be an American and why it was necessary to work to defend America. Work may well be the significant term here, as most issues of the *Magazine* sought to prove that defense was indeed a job, a twenty-four-hour-a-day profession. In so many "Private Snafu" cartoons, and in so many of the newsreels and special features on the *Magazine* bill, work and defense are synonymous, with success on the job equalling success in the war.

Along with trying to shape thinking about the war, then, the magazines also prefigured postwar ideology. We now know, for instance, about such secret sources of wartime power as the Committee for Economic Development, "a coalition of government and businessmen planning for a postwar monopolization of commodity production."[30] But this militarization of private life and the corporatization of the military shared ideological space on the magazines with an insistence on the individual as the true repository of power. In the case of Snafu, his responsibility to keep his weapons clean, or to use his gas mask properly, or to refrain from gossip, posit each recruit as a body politic just as important in deciding the outcome of the war as all of the force, technology, and authority of the state.

The government-produced wartime propaganda was not much different from the Hollywood-produced variety—the feature films and short subjects that played in domestic theaters. Indeed, the very similarity of the mode of exhibition—a Hollywood feature with shorts in both home-front theaters and those in war zones—implies an easy ideological transference from one brand of cinema practice to another. It may well be that it was the film bill itself, along with the subject matter of the films shown, that worked at least partially to convince recruits that the war was worth fighting. Despite the focus in the magazines on war issues and the series' single-mindedness about consensus, the films on the bill probably signified to recruits the same variety and abundance that they had gotten used to by going to the movies at home. This mode of exhibition—short, live-action films on different subjects, a cartoon, and a feature-length film—by so connoting a culture of abundance functioned to mobilize frequently disgruntled recruits to fight for its defense.

4 | THE POPULAR PRESS VIEWS CARTOONS

Shaping Public Opinion While Creating Walt Disney

During the 1940s, such popular magazines as *Time*, *The Saturday Review*, *Commonweal*, and *Popular Science* helped create "Disney Discourse." They developed multiple aesthetic and commercial personas for the filmmaker/entrepreneur, and in so doing, they told their readers not just what to think about the cartoon producer, but about a variety of issues that could be read through him. These journals concentrated on three areas: Disney as a hero of culture and commerce; Disney's perceived decline during the postwar period; and Disney as a corporate boss managing the labor disputes at his studio.

Analyzing these mass media representations of Disney helps us understand how stories about animation in these journals defined cartoons for the public. Indeed, for animation to be written about at all, it had to have been produced by the creator of Mickey Mouse. Max and Dave Fleischer make just a few appearances in magazines during the 1940s and George Pal makes only one or two, while there are hundreds of articles about Disney. Further, and unlike most articles about live-action feature films (which stress aesthetic or narrative issues), the articles about animation are often detailed discussions of filmmaking practice, that is, how cartoons are made. Constructing cartoons as marvels of technology on the one hand and as the handiwork of individual genius on the other, the journals endorsed one of the paradoxes of capitalist mythology: industry becomes a wonderland and work turns into fun, while at the same time workers disappear and the product seems to spring fully formed from the head of the gifted individual.

Popular Journalism in the 1940s

The decade of the 1940s stands out for several reasons as the central era for this analysis. That period marked the maturation of the media empire of Henry Luce, who had founded *Time* magazine in 1923, *Fortune* in 1933, and *Life* in 1936. In addition, the era also signalled Disney's apotheosis as a media hero, dating at least from the 1938 release of his first feature-length cartoon, *Snow White*, and the 1940 releases of *Pinocchio* and *Fantasia*.

Popular journalism flourished during the 1940s in large part because of the steady growth of the middle class. Luce and others like him were responsible for making news a commodity for members of this class and for writing news specifically for them. As a result, even while taking an antilabor position, for instance, journals occasionally exhibited a typically enlightened view toward race relations as long as issues of inequality could be divorced from class. Similarly, magazines of news and commentary engaged in a campaign for public decency directed primarily at depictions of female sexuality, and also, but less obviously, worked for the maintenance of the status quo that had largely created the middle-class audience in the first place.

Between 1940 and 1950, *Time* magazine published more than two dozen articles about cartoons. *The Saturday Review* printed at least ten, *Popular Science* six, *Commonweal* no less than five, and *Business Week* four. These magazines run the gamut of popular magazines from the period, and as a result their articles typify the various forms of Disney discourse. Later in this chapter I will profile the journals themselves. For now it suffices to say that *Time*, like *Life* and *Newsweek*, addressed a huge audience, providing items of general interest about world events, domestic issues, and cultural affairs. *The Saturday Review*, like *The Atlantic* and *Commonweal*, addressed many of these same areas but its pitch was clearly more elevated. *Popular Science* wrote for an audience almost as large as *Time*'s but dealt only with very specialized material—as the masthead indicated, "science and industry." *Business Week* was all business—each of its articles concerned economic issues on a primarily corporate level.

In most of the journals the animation articles appeared in the movie columns. *The Saturday Review* covered cartoons in its "Seeing Things" column in the early 1940s, and then, in the late 1940s, in the "Ideas on

Film" column. *Time* occasionally reviewed cartoons in its weekly "Cinema" section. But both of these journals, along with *Commonweal*, *Life*, and *Newsweek*, also wrote about animation in various other sections. This seems to be different from how feature-length live-action films were dealt with; a look through the magazines shows that these were discussed almost exclusively in the film columns and rarely made it into the "Arts" or "World News" sections. It is both a measure of Disney's special standing and animation's status as a cultural oddity that *Time* would devote columns to Disney in sections labeled "Arts," "Army & Navy," and "Merchandising."[1] *Business Week* discussed cartoons under general news or in the "Labor Management" section, while *Popular Science*, logically enough, did not have a monthly movie review column and talked about animation simply under "News."

Disney, the Great Man

The journals constructed Disney as both a man of the people and a combination corporate magnate and contemporary philosopher. In fact, Disney's image reconciled any number of opposites, not just the proletarian/capitalist dichotomy. In a 1945 article in *The Saturday Review*, Jay Leyda discussed a new book called *The Art of Walt Disney* by Tulane University art historian Robert D. Feild, who had lost an appointment at Harvard in part because of his interest in popular culture (fig. 12). Early in the article, and as a sign of Disney's importance, Leyda exulted that the studio's products covered every subject from "Donald Duck to Beethoven," and then asserted that "here is a medium that unhesitatingly absorbs the artistic achievements of the past and present in order to entertain with success and to devote unlimited effort to future experiment."[2] In just one sentence Leyda used Disney's cartoons to smooth

Figure 12. Walt Disney on the cover of *The Saturday Review of Literature*, in which Jay Leyda reviewed Robert D. Feild's *The Art of Walt Disney* (6 June 1942).

RSE CAVE, KY., BUYS A BOOK, By Theodore English

The Saturday Review of Literature

Vol. XXV No. 23 New York, Saturday, June 6, 1942 Fifteen Cents

Walt Disney (above) gave Robert Feild free run of his studio and out of it came a "brilliant and satisfying analysis." (See Jay Leyda's review of "The Art of Walt Disney," page 5.)

over one opposition after another: Donald Duck—low art—and Beethoven—high art; the best in art from the past and the present; contemporary popular approval ("to entertain with success") and an unwavering commitment to the new and untried ("unlimited effort to future experiment").

Then Leyda applauded Feild's ability to explain "the delicate balance between the creating artist and the huge mechanism of the Studio," with Disney here reconciling art and commerce in a way that seems to bring added credit to both.[3] Disney, then, bridged gaps between areas that were—and still are—often perceived as unbridgeable: art and commerce, high art and low art, immediate acceptance and trailblazing the future.

Time constructed Disney as a hero of reconciliation in a 1945 article called "Mickey's Coworkers." The magazine talked about the studio's planned feature-length cartoons and said, "to give voices to his furred and feathered folk, Impresario Disney signed up . . . Dinah Shore, the Andrews Sisters, Edgar Bergen. To supply the cartooned creatures with plots and dialogue, he has engaged such litterateurs as . . . Playwrights Marc Connelly and Edwin Justus Mayer, Author George Rippey Stewart, Author-Critic Sterling North and Folklorist Carl Carmer." *Time* insisted that "few dreamers dared hope for a postwar caterpillar with the voice of Nelson Eddy and the brains of Aldous Huxley."[4] It was a sign of Disney's greatness, however, that he would do just that; that is, combine the most popular baritone of his time—Eddy—with the chronicler of it—Huxley—and mix popular entertainment with high art in ways unimaginable even to "dreamers."

Throughout Disney's honeymoon with the press, his powers of reconciliation were deemed to go beyond mere filmmaking. In a wartime series of *Saturday Review* pieces, film producer Walter Wanger explained how Disney was able to bring together all of the branches of the armed forces, the Cabinet departments, and even a foreign government or two. "Hollywood," Wanger said, was "as busy as a league of nations," and it was the Disney studio and, in fact, Disney himself, that best typified the league's goals of unification: "The one and only Walt is carrying on with the Army, the Navy, the Department of Agriculture, the Office of Coordinator of Inter-American Affairs, the Canadian Government, etc."[5]

Finally this kind of smoothing over of oppositions, this creation of a place where branches of government as well as governments themselves came together, got back to making movies. After discussing the presence of "an American Naval Officer who has just returned from experiences on two carriers" and, "at another table . . . [the] British officer, sent here by the English Government," Wanger talked about Disney's government propaganda films and said, "all of this . . . is coming from the modernistic pastel-colored fairyland plant where thousands are creating and delivering in the interest of the new education of the free world." With a war in full swing, that which critics often considered the essence of animation—its ability to produce a fantasy world out of paint and ink—was extended to the studio itself, Disney's "fairyland," which was also, in keeping with a wartime emphasis on production, a "plant." Furthermore, the primary product of this fairyland had become propaganda, that which would educate the free world. Rather than remarking on the incongruity between make-believe and mass persuasion, Wanger celebrated it. About the feature-length *Saludos Amigos*, "the result of the Disney-Rockefeller expedition to Latin America last year," Wanger exulted, "Watch for it. It is really something!"[6]

Disney's Decline

Just a few years later Disney began losing favor with the critics, at least in part because they no longer believed in his powers of reconciliation. In a respectful if, finally, unfavorable review of *The Three Caballeros*, a 1945 feature-length cartoon, *The Saturday Review* looked at Disney historically and claimed that, earlier, his "genius" had been in appealing "to what is childish in adults and adult in children." Besides turning a splintered audience into a unified one, Disney also merged his art with large-scale production and diversification, and the review approved of this, too. "What with his books, his records, his gadgets, and his toys," the review claimed, "no less than with his films, there was scarcely a nursery he had not invaded or a small mind he did not occupy."[7]

"Invaded" here was intended to be taken benignly; almost all of the journals talked about Disney's imperialization of the home, and they did so uncritically. The reviewer had no complaint with Disney's role as part Hans Christian Andersen, part corporate conglomerate. For critics, what truly marked Disney's decline was his inability to bring together the disparate groups in the movie audience. *The Saturday Review* could now say that, even as early as *Snow White*, which a few years before the journals would not disparage, "The Witch . . . in the excess of the terror she created . . . made clear that Mr. Disney as a teller of fairy tales was having audience trouble; that he could not quite make up his mind whether he was appealing to the young or the old; and that he was trying both to have his cake and eat it."[8]

Disney's perceived inability to appeal across audience divisions demonstrated just one of the cartoon producer's problems. As noted earlier, in the same year as this review of *Caballeros*, *Time* had praised Disney's proposal to merge high art with low, Nelson Eddy with Aldous Huxley. But in the end those plans fell far short of expectations. *The Saturday Review* complained in its *Caballeros* review that even in *Fantasia* great music was sometimes visualized "in terms of the most cloying Christmas cards." The magazine then pointed out that Disney's trademark "wonderful touches" were offset by the "bad features" which were "violently inartistic . . . cheaply Hollywood." Walter Wanger's laudatory "league of nations" had changed. Disney was still "an Ambassador of Good Will" but now he was marked by diffuseness rather than unification. "The portfolio he carries," remarked the reviewer, "turns out to be a strange grab-bag."[9]

During this period of Disney's critical decline, which dated from about 1945, *Time*, like *The Saturday Review*, still discussed Disney in terms of oppositions, but now they were usually irreconcilable. In a generally disapproving 1948 review of a new feature, *Melody Time*, the magazine began by reciting the film's pedigree, its status as a part long-hair, part low-brow hybrid, featuring sequences with "Rimsky-Korsakov's bumblebee" and also Pecos Bill. But by now "nearly every attempt at cuteness, sweetness, tenderness, sublimity, results in one or another kind of painful simper. . . . On the other hand, the straight technical expertism is still one of the wonders of the movie world."[10] The animator who used to bridge art and commerce, popular culture and serious literature, now could not even make a unified film. The "technical expertism" which was a sign of Disney's genius (and which belonged almost solely to Disney even as a critical term; *Time* talked about few other filmmakers' technique) now only seemed to call attention to the weaknesses of story and character.

In a convenient bit of movie release-date timing, *The Lady from Shanghai* was reviewed in the same issue as *Melody Time*, and Disney's role as reconciler-in-chief fell to that film's writer/director, Orson Welles. Determined to criticize or praise primarily in terms of bringing elements of a movie together, the reviewer noted that "the big trick in this picture was to divert a head-on collision of at least six plots, and make of it a smooth-flowing, six-lane whodunit. Orson brings the trick off."[11] A critical discourse that grew up largely around Disney now was used against him, and to celebrate a younger and very different kind of filmmaker.

Critical Shifts and Product Transformation

A brief description of Disney's production history during the period is pertinent here. When Disney began making feature films he apparently chose subjects and genres very carefully. To maximize profits he needed to differentiate his product not only from that of other cartoon producers but from live-action feature filmmakers as well. He also had to guarantee, however, that his films were not too different so as not to estrange a large portion of his potential audience.

For his third feature, which appeared in 1940, Disney made *Fantasia*. Looking like no other full-length film—it combined classical music with animation in a series of short sequences, with an appearance from high culture icon Leopold Stokowski—*Fantasia* did not seem that dissimilar from Disney's earlier, shorter *Silly Symphonies*. Still, *Fantasia* was more or less one of a kind, because for his first two features and also for his fourth and fifth, Disney animated familiar fairy tales—*Snow White* and *Pinocchio*—and created new ones—*Bambi* and *Dumbo*. Live-action sound features had rarely been fairy tales before *Snow White* (*Alice in Wonderland*, from 1933, comes to mind as one of the few) but the narrative for *Snow White*, stressing the elimination of the obstacles faced by the romantic heterosexual couple, hardly differed from that of many Hollywood films. Moreover, the fairy tale had been a staple of cartoon production for a number of years. Disney, therefore, in just a few years, had managed to establish himself as an important artist with *Fantasia*, had carved out his own niche with his production of fairy-tale features, and had also managed to remain completely within the mainstream of Hollywood production.

During the war Disney switched genres. His features were either didactic—*Victory Through Air Power*—or somewhat removed from the fairy tale because of the entrance of live-action footage and identifiably contemporary settings—*Saludos Amigos* and *The Three Caballeros*. At least through *Saludos Amigos* critics were enthusiastic about Disney's wartime production, features as well as shorts. They viewed these films as valuable propaganda and as proof of Disney's status as an unofficial ambassador of good will in general and Americanism in particular.

Because of the critics' conviction of Disney's significance and stature, they couched their postwar criticism in the most cautious terms. John Mason Brown in *The Saturday Review* typified the journals' attitude during Disney's perceived decline. He practically apologized for disliking *Cinderella* and was more than ready to blame himself rather than the filmmaker: "But to return to Mr. Disney and my present indifference to an erstwhile enthusiasm. Is it because I have grown older? Is it because repetition is bound to have robbed his work of its novelty? Is it because Mr. Disney at present fails me or I fail him? Or is it because of the very nature of the medium at which he excels? Is it because this is a medium that, in spite of the courage and success of his attempts to alter and enlarge it, is nonetheless limited?"[12]

Brown thus apparently assumed that, because cartooning was not the dominant filmmaking practice (live-action feature-filmmaking was and still is), it must be "limited." But Brown was asking interesting, demystifying questions as well as absolutely naive ones. Furthermore, the issues he brought up, rather than being personal ones (about his own changes and about Disney's courageous attempts to alter the medium) were instead broader ideological ones. Disney's critical decline came about because of changing cultural attitudes toward such disparate subjects as genre, cartoon style, audience, representations of female sexuality and race, and the development of television. It is significant, too, that one aspect of the Disney empire would not be criticized: its entrepreneurship.

Animation, Television, and Realism

Commercial television had been put on hold during the war, but in the early 1940s there could be no doubt that the new medium would be developed, and the motion picture studios actively sought to guarantee their control over it.[13] Disney had been preparing for television at least since 1936, when he left his distributor, United Artists, because it would not give him future television rights to his cartoons.[14] Now, several years later, Disney would have to go through the process of differentiating his product from the competition produced by television.

It seems likely that Disney, sensing the kind of entertainment product that television might provide, would develop a cartoon form that the new medium, for technological as well as economic reasons, could not reproduce: the combined live-action/animated film. It is also probable that Disney would reconsider, at least somewhat, the future of feature cartoon production. Not only was a feature expensive, it also was not clear in the 1940s that the full-length cartoon was the form of animation best suited to television. Perhaps at least in part to save money, but also to prepare as fully as possible for television, Disney released *Fun and Fancy Free*, *Make Mine Music*, and *Ichabod and Mr. Toad*, late 1940s features that, like *Fantasia*, were really collections of shorts, which could easily be broken down for television presentation.

The journals, however, viewed television not so much as that which would bring permutations to movie forms but as a potentially frightening invasion of the home, something that would have an unspecified but probably unhealthy effect on children.[15] As a result, the journals did not examine Disney's changes as a response to a new medium. Instead, those changes remained, simply, inexplicable and somewhat unsettling, much like the new technology itself. For example, in discussing *The Three Caballeros*, the first of the Disney features to receive generally unfavorable reviews, John Mason Brown in *The Saturday Review* could only wonder "why Mr. Disney and his staff . . . should have tried to mix human beings who are only real with drawn figures who are pure fantasy." Brown added that "compared to this unhappy mixture, the home life of bromo and seltzer is bliss itself."[16] A few years later another *Saturday Review* article asserted that "Mr. Disney's taste has always been an uncertain factor. . . . Certainly his mixture of drawn and real people . . . was one of the most unfortunate experiments since Prohibition."[17]

Similarly, *Time* called *Song of the South* (1946) "a curious mixture of live action (70%) and cartooning (30%)," and lamented that there was not "a much heavier helping of cartooning."[18] A year later, in 1947, *Time* seemed tired of the combination, beginning its review of *Fun and Fancy Free* by informing its readers that the film "is *another* Disney movie that mixes cartoons and live actors [emphasis mine]." The review added that "in spite of the Disney technical skill, it has never been a very good idea to mix cartoons and live actors."[19] In 1950, in a fairly positive review of *Cinderella*, the unnamed *Time* critic assured readers early on that the film was "unalloyed make-believe, without the disenchanting sight of a single photographed human face."[20]

Even animated humans, drawn and painted rather than "live actors," were grounds for disapproval by the critics. *Cinderella* disappointed the film reviewer for *The Saturday Review* in part because the cartoon presented "an unusually large number of human characters."[21] *Time* faulted the same film because the main characters—the prince, the witch, and "the blonde heroine"—"are drawn in an attempt at literal likeness that the best technique of animation never brings off without a certain stiffness."[22] Both magazines disliked Disney's new preoccupation with humans rather than animals, feeling that it provided a kind of realism that should be foreign to animation. Thus, having established animation as the domain primarily of animals and always of drawn figures rather than actual performers, Disney was punished by the journals for violating his own rules. Instead of appreciating Disney's genre shifting (combining the travelogue with the cartoon in *Saludos Amigos*, for instance), and his attempts at pastiche (mixing live action with animation), the critics could only shake their heads at the animator's transgressions.

Disney's decline in the journals, then, was due at least in part to the inclusion of more realism—both drawn and live-action—in his films. In 1942, Jay Leyda worried in *The Saturday Review* that "the unceasing struggle of the whole [Disney] organization to provide vivid outlets for unbound imagination [might] bind itself unavoidably to a single precise style or a single habitual attitude.[23] Indeed, by the late 1940s, Disney's "imagination" had allowed him to broaden his style considerably with the combination of live action and animation. Leyda's worries about artistic stagnation were unfounded, but the critics, nostalgic for the old formulas and styles, were unmoved, except to wish that Disney had stuck to his "single habitual attitude."

Disney Cartoons and the Response to Racism

When, in at least one respect, Disney maintained that "attitude," many of these same critics begin to question him. During the postwar period, it became fashionable to scrutinize that which had been a staple of Disney cartoons for many years; their racism. Through 1945 and *The Three Caballeros*, this issue never came up in any of the journals I examined for this book. In a 1941 review *Time* could say that the "five black crows are to *Dumbo* what the Seven Dwarfs were to *Snow White*. Their burlesque song-&-dance routine, hilarious, eminently crowish, is typical of the good circus humor that bubbles through the picture. They have one of the best ('When I See an Elephant Fly') of *Dumbo*'s nine tuneful melodies."[24]

Just as there was complete acceptance of these animated minstrel show characters and no mention that they displayed all of the racist stereotypical behavior associated with black people, Disney's depictions of South Americans were also discussed uncritically. Indeed, reviewers accepted *Saludos Amigos* and *The Three Caballeros* simply as proof of Disney's Good Neighborliness.

Following the war, however, and at least in *Time* magazine, race could be mentioned in film articles and racism could be criticized. Just five years after the endorsement of the crows in *Dumbo* and four years after a review of the George Pal Puppetoon *Jasper and the Watermelons*,[25] in which Pal's Stepin Fetchit–style stereotyping received no mention, *Time* began a remarkable series of Disney critiques.

In its 1946 review of *Make Mine Music*,[26] *Time* said that "the hillbilly ballad *The Martins and the Coys* is a burlesque of backwoods feuding which will delight lovers of radio rurality and of Paul Webbs's mountaineer cartoons." The magazine then cautioned, however, that this segment of the Disney film would "offend those who think such caricature as insulting as the hush-mah-mouf kind of comic contempt for Negroes." Here, a Disney depiction of a white subculture—hillbillies—was likened to the kind of blackface routine that one finds in *Dumbo*. For the first time in *Time* cartoon reviews, race became an issue and black stereotypes were acknowledged and labeled as offensive. There was, of course, something like fence-straddling in this discourse, as the reviewer first praised the sequence and then claimed that the depiction of the hillbillies was potentially offensive only to those who were sensitive to such things. The reviewer avoided expressing a definitive personal opinion about the cartoon.

Just a few months later, Disney released *Song of the South*, his version of Joel Chandler Harris's stories. This film had a central black character, Uncle Remus, and *Time* quickly got to the issue of race. In the second paragraph the reviewer wrote that "tattered ol' Uncle Remus, who cheerfully 'knew his place' in the easygoing world of late 19th Century Georgia (Author Harris, in accepted Southern fashion, always omitted the capital from the word 'Negro'), is a character bound to enrage all educated Negroes, and a number of damyankees."[27]

Making this mention of race politically safe to a middle-class readership was the assertion of the benefits of middle-classness. *Time* assured its readers that a black audience's sensitivity to racial issues was in fact a function of class privilege, as only "educated" blacks would take offense. Poor blacks, apparently, would not be bothered. Still, this review shows how distinctly the discourse about Disney had changed and how politicized that discourse had become in *Time*. Racism was not only an issue to be discussed, but to be discussed almost immediately, here in the second paragraph. While Disney had once been written about as a reconciler, as one who brought opposites together, he was now viewed as a force that worked to polarize. The review pointed out that the film would be viewed differently in the North and South and even from race to race, as it was blacks who were singled out as the viewers most likely to be upset. In just a few years, Disney had gone from global good-will ambassador to a symbol of domestic divisiveness.

The acknowledgment of racism in *Time* shows the complexity of positioning a magazine ideologically. Like the other journals, *Time* frowned on Disney's experimentation with combining live action with animation and, as I will show, sided with management over labor during worker disputes at the Disney studio. Yet the magazine was able to recognize racism during the late 1940s and take some offense at it, and to describe the country, as it did in the review of *Song of the South*, as divided over attitudes about race.

This small step in the liberalization of a mass-market magazine occurred at a time when racism and civil rights had become mainstream subjects in *Time*'s news sections as well. In 1947, for instance, that mythological repository of American values, major league baseball, began allowing blacks to play for the first time in the twentieth century. Also, a war fought at least in part against prejudice (and in which black troops played a prominent role) may have forced those who shaped editorial policy at the journals to come to grips with American prejudice at home; South America was fair game for caricatures (the journals approved of José Carioca in *Saludos Amigos*) but representations of black citizens of the United States had to be judged against standards that had been ignored by most white Americans just a few years before.

As a result there were different kinds of films being made by the major studios and different critical attitudes developing toward them. During the war, as the reviews of Disney's *Saludos Amigos* demonstrate, critics approved of the cinema as a propaganda machine. After the war, however, with the increased activity of the House Un-American Activities Committee, the connotations of propaganda changed. Films could still be expected to develop a political discourse, but one that was humanist rather than propagandist. For instance, such major studio films as *Gentleman's Agreement* (1947) and *Crossfire* (1947) would be applauded by critics for examining anti-Semitism in the postwar era. In addition, foreign films, which had been deprived of wide distribution before 1945, were now in more American theaters because the Justice Department's Paramount Decree weakened the major film companies' control over theatrical exhibition. Some of the more celebrated of those films, such as the neo-realist *Open City* or *The Bicycle Thief*, helped to make it more fashionable for the middle-brow journals to discuss, however timidly, political issues.

Even in this context of changing attitudes about race and the depiction of social problems in movies, some of the journals remained unaffected. An intellectual journal such as *The Saturday Review* refused to acknowledge that race might be an issue in a Disney film even when its reviewer was criticizing the Disney product. In his 1950 analysis of *Cinderella*, for instance, which also surveyed Disney's career over the last dozen years, John Mason Brown celebrated that, immediately after the release of *Snow White*, "Mr. Disney received an honorary degree at Yale . . . which described him as the 'creator of a new language of art, who has brought the joy of deep laughter to millions . . . without distinction of race.'"[28]

For Brown, those were clearly Disney's salad days, but his decline had nothing to do with increasing audience sensitivity to the racism in his films or to a questioning of whether Disney's films really functioned "without distinction of race." Disappointed in *Cinderella*, Brown still saw something of the old Disney in the song "Bibbidi Bobbidi Boo," and wrote that "I must admit I rejoiced in something of that same sense of release which all of us used to experience when we responded to 'Who's Afraid of the Big Bad Wolf?' 'Whistle While You Work' . . . or 'When I See an Elephant Fly.'"[29] Of course, that last song, performed by the black crows in *Dumbo*, was the one that most obviously stereotyped black behavior. Even by 1950, however, Brown proved himself incapable of recognizing that racism; instead, *Dumbo* was the object of his nostalgic longing for a period when the quality of a Disney film could not be questioned. Priding itself on its examination of art and culture, *The Saturday Review*, long after *Time* had changed its position, still viewed Disney as an artist whose work must be judged on primarily aesthetic grounds.

Popular Science, with concerns and an audience quite different from those of *The Saturday Review*, nonetheless also failed to discuss the politics of race. During the war the magazine fully approved of Disney's *Saludos Amigos* and *The Three Caballeros* and accepted cartoon caricatures as honest depictions of South American nationalities. In an article about the technology behind live-action/animation combinations in *Caballeros*, *Popular Science* claimed "up goes another character in the Walt Disney Hall of Fame . . . while José Carioca, the parrot of *Saludos Amigos*, typified Brazil, or the Portuguese . . . Panchito leans definitely toward the Spanish side."[30]

During the latter half of the 1940s the magazine's only articles concerning animation did not treat Disney at all. Instead, they dealt with two George Pal imitators and with Pal himself. Writing about a three-dimensional wax doll cartoon by Larry Morey and John Sutherland, *Popular Science* admired the two main characters, Lolita and Pepito. Describing the wax-pouring process Morey and Sutherland used to make their dolls, the magazine said that after "two hours . . . they are ready to be worked into the desired shapes and forms that make them look like real people."[31] As the photographs in the magazine demonstrated, the dolls did not look real at all. However, probably because they were supposed to be Mexican, the dolls could be considered real in the same manner that José Carioca could be considered typical of Brazil.

A few months later *Popular Science* examined one of George Pal's more extraordinary films, *John Henry*. This "Puppetoon" about black railroad laborers subscribed to most of the stereotypes of Pal's *Jasper* series, although *John Henry* also detailed the effects of industrialization on the black working class. Determined to celebrate technology, however, the magazine stressed the fairy-tale aspects of the film instead of the social issues that it raised. In the first paragraph, the article stated that "John Henry was the fabulous super-strong man of the workers who built the railroads," and that his "great contest against the machine is an epic of American folklore."[32] In so saying, *Popular Science* had safely placed the cartoon within the genre that the journals always favored for cartoons—the folk/fairy tale—and mentioned nothing about John Henry's race or the catalogue of stereotypes that formed his character.

"Erotomaniacal Regard"

That *Time*, by far the most widely circulated of all of the journals under consideration, could discuss race in relation to cartoons, and specifically in relation to a cultural icon of Disney's magnitude, indicates the extent to which an examination of prejudice had entered popular discourse. Other journals, though, comfortably refused to follow *Time*'s lead. Yet on a different issue, and one that also was used in the latter half of the decade to criticize Disney, the journals were able to reach an agreement. They disapproved of what they viewed as a new eroticism in Disney's cartoons.

This new interest in the issue was at least as much a function of Disney's experiments in animation as it was a sign of changing social mores that suddenly made it possible to speak about sexuality. With his combination of live action and animation and with the perceived new realism of his human cartoon characters, Disney had not so much changed the attitudes in his cartoons toward sexuality as he had changed his representation of it. Instead of anthropomorphic flowers or animals there were now real-life (that is, live-action) women in such films as *Saludos Amigos* and *The Three Caballeros* and drawn women, like Cinderella, who suddenly seemed more human to the critics.

Disney's increased realism made the critics notice for the first time how the cartoon producer presented sexuality. Writing about *Caballeros*, the *Time* reviewer complained that "thanks to an ingenious but seldom very rewarding blend of drawings and regular color-movies with living actors, Donald [Duck] whizzes from one Latin American beauty to the next like a berserk bumblebee." The critic continued that "since he remains at base a combination of loud little boy and loud little duck, his erotomaniacal regard for these full-blown young ladies is of strictly pathological interest."[33]

It was not simply adolescent animals in pursuit of women that offended the critics. A year after the review of *Caballeros*, *Time* examined the Disney compilation cartoon, *Make Mine Music*. The magazine appreciated some of the sequences but described "All the Cats Join In" as "a jukebox setting of Benny Goodman's record, in which orgiastic hepcats and bobby-soxers, mad on chocolate malteds, tear all over the place."[34] "Erotomaniacal" and "orgiastic" were new words in the critical discourse about Disney, but the sequences they described, as I discussed in chapter 1, were hardly new to cartoons. Because the look of the cartoons was different, however (they combined live action with cartoons), or at least was perceived to be different (characters seemed more real to the critics), the reviewer of a Disney film felt responsible during the late 1940s to alert viewers to representations of sexuality and to criticize Disney for his excesses.

The Saturday Review employed more reserved terminology but nonetheless was in agreement with *Time*. About one of the sequences in *The Three Caballeros* that combined live action with animation, John Mason Brown said simply that "it is impossible to understand why Mr. Disney and his staff should ever have sent their caballeros on that magic serapi ride across a beach full of bathing beauties"[35] (fig. 13). Five years later, in 1950, Brown complained that "Mr. Disney's Cinderella . . . is a blank-faced blonde, armed Al Capp-a-pie with the allurements of Daisy Mae. . . . There is nothing of the ill-used waif about her."[36] Snow White clearly was Brown's prototypical waif, and she was hardly less real than Cinderella. For Brown, then, even though he disapproved of the new realism in Disney's films, the problem with Cinderella was not merely that she looked too human but that she appeared too mature, and therefore more sexually suggestive than a cartoon heroine ought to be.

Disney's advertising strategies were at least partly to blame for the critics' reactions to his films. For *The Three Caballeros*, his publicity material stressed not only the combination of live action and animation but the combination of animated animals with real women. Accompanying the *Time* review of the film was a photograph, undoubtedly supplied by the Disney studio, which the magazine captioned "Donald Duck & Friend."[37] It showed one of those "Latin American beauties" who was the object of the duck's "pathological interest," with Donald standing next to her. She smiles at him and rubs a finger under his bill, while Donald, obviously smitten, smiles back, eyes closed and hands clenched over his heart. Under the caption the magazine added an opinion about the couple, calling them "an alarmingly incongruous spectacle." What for Disney must have seemed like a sensible way of advertising the film—a publicity photograph that stressed the live action/cartoon combination—was instead used by *Time* to criticize it because of the way the photo also stressed sex.

In the same *Saturday Review* that discussed *The Three Caballeros*, an advertisement appeared for the film. It showed none of the animated characters but only a woman in a bathing suit, hand poised on hip in typical bathing beauty fashion. The ad copy exclaimed "Yes! She's real! Alive and lovely in a Walt Disney Picture! It's amazing, wonderful and thrilling!"[38] With this advertisement Disney assured his audience that his new film contained no waifs, and from the ad alone it would be difficult to be certain that the film was a cartoon at all (fig. 14). Disney chose to emphasize not just the live action in the movie but more specifically the live-action women, a tactic that, in the journal reviews at least, backfired.

Beginning with the first cartoon produced specifically for television, 1949's *Crusader Rabbit*, animation would be ghettoized as children's entertainment on television, and the journals told their readers more and

Figure 13. "It is impossible to understand why Mr. Disney . . . should have tried to mix human beings who are only real with drawn figures who are pure fantasy." A publicity still from *The Three Caballeros* (*The Saturday Review*, 24 February 1945).

Figure 14. Advertisement appearing in *The Saturday Review*, 24 February 1945.

more during the late 1940s that theatrical cartoons were more fit for children than adults. Rather than giving in to this diminution of his audience, Disney might have believed that the best way to expand his viewership was to stress the adult aspects of his movies.

In fact, in its review of *Make Mine Music*, the 1946 cartoon with the "orgiastic" bobby-soxers, *Time* stated in the first sentence that this "poor man's *Fantasia*" was "prepared for the 18–27-year-old age group which has heretofore proved least responsive to Disney films." Five years earlier *Time* reported Gallup Poll findings that the "ace cinemaddicts are 19-year-olds. . . . Laggard are those over 30. . . . Typical movie-goer is 27 years old."[39] Although never relying heavily on those viewers over twenty-seven, Disney may have felt a real pinch if he was losing an audience of people over eighteen. The entrance into the cartoons of live-action women and sexualized bobby-soxers may have been Disney's last gasp before giving in to the audience shift in animation and devising ways to exploit it to the utmost.

By the end of the decade Disney had stopped mixing live action with animation, and his feature-length cartoons during the 1950s—*Alice in Wonderland*, for instance, or *Lady and the Tramp*—returned to the fairy-tale motifs that he had partially abandoned. Disney also gave himself over fully to children's entertainment, not only in his theatrical cartoons but in his major commercial achievements—his television program and his amusement park. Disney seems to have taken the critics to heart, using their judgments about audience and proper cartoon subject matter and style to increase his popular culture empire.

Entrepreneurial Heroism

Roughly from 1945 to 1950, the movie columns in *Time* routinely disparaged Disney's films. On its business pages, however, Disney remained the model corporate boss. In a 1948 column headed "Merchandising," *Time* ran an article called "The Mighty Mouse."[40] "In Burbank, Calif. last week," the article began, "Walt Disney was presented with the 5,000,000th watch to be manufactured with the name and likeness of Mickey Mouse. As he put it among his trophies, Disney smiled at a beaming, moon-faced Manhattan salesman named Kay Kamen." The

photograph accompanying the article shows Disney, with Kamen at his side, smiling widely as he reads their new contract; in both text and picture, Disney was the happy, rightly rewarded entrepreneur (fig. 15).

Disney smiled not only because of all of the watches that had already been made but because of the products still to come off the assembly line. The second paragraph of the article told readers that "Disney and Kamen had just signed a seven-year renewal of a 16-year business association." Commercial continuity had become practically familial as business relations lasted for years, with *Time* also detailing the success of the partnership: "This year goods bearing the faces of Disney characters will bring in a retail gross of $100 million."

To explain just how wide an area $100 million worth of products covered, the article pointed out that Kamen "defines his sales territory as reaching 'from the Isthmus of Panama to Hudson Bay,'" and added that "after Jan. 1 Disney's brother, Roy, will handle the rest of the world." Once again, entrepreneurial discourse became familial discourse as the magazine stressed the role of brother Roy. *Time* also celebrated Disney's brand of cultural imperialism—his empire now included "the world." During the war the journals made much of Disney's Good Neighbor policy toward South America. *Saludos Amigos*, for instance, was considered excellent propaganda for a united Western hemisphere dominated by United States capital. With the war over, interest shifted, however, from an emphasis on Disney as disseminator of U.S. political programs to Disney as kingpin in a world market system.

In order to make this kind of dominance seem as far removed from acquisitive capitalism as possible, *Time* discussed it not only in familial terms but also as a sort of American-dream-come-true for Kamen. *Time* explained that Kamen "considers himself the world's greatest Disney fan. Whenever a new picture is completed, he flies to Hollywood to preview it, begins selling its characters before the film is even released." For Kamen, as he was interpreted by *Time*, selling Disney was simply a byproduct of loving Disney. He flew to Hollywood to preview Disney films primarily because he could not wait to see them. *Time* implied that

Figure 15. "Kay Kamen . . . salesman of exclusive manufacturing rights for Mickey Mouse and other Disney characters . . . defines his sales territory as reaching 'from the Isthmus of Panama to Hudson Bay'" (*Time* magazine, 25 October 1948).

Kamen would not be a pitchman for a filmmaker whose films he did not like.

Time then provided a partial Disney creation myth, saying that "Disney and Kamen both started their early careers in Kansas City. There, in the early 20's, Disney was struggling along as a commercial artist, Kamen as a sales promotion man for department stores." In the 1930s, "Kamen went out to Hollywood, sold him [Disney] the idea of letting him handle the manufacturing royalties. This was an aspect Disney had neglected." Each man functioned as a complement to the other; Disney, neglecting business, provided the art, and Kamen provided the know-how, with Kamen himself winning *Time*'s implicit admiration for taking the initiative to approach Disney. Thus, in the photograph and according to the article, Disney and Kamen smiled at the new contract not simply because it promised immense profits but because they themselves were like brothers, because they had experienced, and triumphed over, hard times, and because their domination of world markets appeared to be benign and beneficial.

The Saturday Review was less likely than *Time* to notice, let alone celebrate, Disney's business contracts. However, the journal was careful to point out that there was absolutely nothing wrong with Disney making as much money as possible. In a 1950 issue of the *Review*, anthropologist Hortense Powdermaker wrote "Celluloid Civilization," in which she analyzed the conflict between art and commerce in Hollywood. After criticizing those writers, producers, and directors who cultivated mediocrity in their films, Powdermaker was quick to point out that "some of our most creative popular artists, such as Chaplin, Gershwin, Walt Disney, and Irving Berlin, have made fortunes," and that "the artist can contribute to business."[41] Powdermaker's assertion stands out as a welcome debunking of the romantic assumption that all creative people must suffer in poverty. But she also claimed that Disney himself was an artist rather than an entrepreneur who employed some artists, and that his fortune was a just reward for his artistry. Powdermaker explained that Disney ran his business primarily to create art rather than to accumulate profits, or at least to exercise his creativity and make money in equal doses: "The real artist in Hollywood cannot be completely satisfied, even though he earns a fortune, if he is not functioning as an artist."[42]

It is at least a slight measure of the journals' belief in Disney's postwar decline that *Popular Science*, which printed three articles about Disney between 1942 and 1944, did not mention him again until 1953. At that time Disney the cartoonist was tangential to Disney the entrepreneur; the article, called "Walt Disney Builds Half-Pint History,"[43] referred to him as "the celebrated creator of Mickey Mouse," but dealt primarily with the creation of Disneyland. The article admitted that "Disneyland . . . still has a long way to grow," and then added that "its purpose is to entertain people of all ages and also to teach them by means of tiny but exact models how life in the U.S. developed to its present level." This Disney was all business, as the article detailed one of his primary commercial ventures (the other being his move to television) during the 1950s.

Although *The Saturday Review* and *Popular Science* had few interests in common, their projects in relation to Disney during this period were very similar. Like the Powdermaker piece in *The Saturday Review*, the *Popular Science* article applauded Disney's entrepreneurship and attempted to link it to values not usually associated with high-powered capitalism. For *Popular Science*, Disney's business enterprises were noteworthy not so much because they were tempered by art, which was Powdermaker's reasoning, but because their primary function was education, that is, teaching people of all ages.

Whistling While They Work: The Journals and Labor

In a 1942 issue of *Time* magazine, Walt Disney asked a question about *Snow White* and then answered it: "Do you know how long it would have taken one man to make that picture? I figured it out—just 250 years."[44] Other articles from the same period provided similar information about the armies of people necessary to produce the feature-length cartoon *Snow White* or even a *Mickey Mouse* short. At least by the 1940s, the general public knew full well that Walt Disney–style animation was the most labor-intensive form of Hollywood film production.

Still, those same journals just as frequently made the labor process disappear from animation. In a 1943 *Saturday Review* article called "Mickey Icarus," for instance, Walter Wanger described Disney's new cartoon, *Victory Through Air Power*, by saying that "every once in a while a motion picture flashes across the horizon to prove our industry an instrumentality of human enlightenment."[45] This was the same Wanger who was one of Hollywood's most successful producers, but his assertion implied that good films came virtually from nowhere, more similar to natural phenomena than to corporate artifacts.

The 1940s stands out as an especially interesting decade for studying this apparently contradictory popular discourse about labor at the Walt Disney studio, because the decade witnessed the spring-through-summer 1941 strike there. About 40 percent of Disney's workers—over four-hundred employees—either joined the picket line around the studio or honored it, with the major differences between labor and management concerning wages, screen credit (only Disney's name appeared on the studio's cartoons), and the right to unionize.[46] Most of the strikers were lower-echelon cartoonists—assistant animators, inkers, and painters, for instance—although they were joined by a few top animators. At least through the end of the 1940s Disney remained rancorous toward the employees who had participated in the action against the studio. For all of Disney's artistic milestones and commercial successes, therefore, which the journals from the decade duly noted, there was also the producer's deteriorating relationship with labor, which the journals mentioned only occasionally. By looking at how the journals from this period talked about labor at the Disney studio, we can at least partially understand what millions of people—the readers of these magazines—learned about artistic creation and factory production and about the equilibrium or imbalance between the two.

Labor and the Liberal Press

My examination of animation labor issues in the journals begins with *Commonweal*, which was founded in 1924 and called itself a "weekly review of literature, the arts, and public affairs." Although it achieved a readership that went beyond the Church, the magazine mainly attracted an audience of mildly liberal Catholics.[47] George Wolseley, in his study of popular journalism, placed *Commonweal* in his high-class group with other magazines directed at "the educationally and economically higher or upper-income citizens." The magazine appears to have succeeded in attracting a very specific audience, as Wolseley pointed out that "from the popular viewpoint," which for him seems to mean those who were not economically or educationally privileged, *Commonweal* and magazines like it were shunned for their "essential dullness." Although speaking to an elite audience, *Commonweal* defined itself in a twenty-fifth anniversary editorial as "working essentially for the extension of social justice and for the protection of the human person in a period when he is threatened by impersonal, inhuman bigness, by machine-like institutions of government, industry, education, and propaganda."[48]

This of course is the classic liberal platform: a commitment to a system of justice that poses no threat to the class system and a prioritizing of the individual—the "human person"—not just over potentially oppressive institutions, but also over the collective. Unlike the other journals, then, when *Commonweal* made references to the strike at the cartoon studio it was willing to blame Disney at least partially for some of the problems.

In an equivocal 1941 review of *The Reluctant Dragon*, a collection of shorts strung together to make a feature, Philip Hartung discussed the premise that connects the cartoons. "The whole," he explained, "is really a tour through the Disney studios. The tourist is our old friend Robert Benchley, who shows . . . that education can be screwy fun under the proper tutor." Benchley "stumbles into the various departments and learns what goes on in each one. He sees drawings made, sound effects added, the rainbow room where the beoootiful colors are mixed. . . . Donald Duck makes an appearance. And Goofy gives a riding lesson." The film presented the same view of the studio found in most journals. It was a place where one could only marvel at the art being made, and where one felt the almost tangible presence of the various cartoon characters.

Hartung pointed out, however, that the "one thing that Bob does not learn about as he sees the beautiful and orderly Disney studio is the strike in evidence there now. Of course, Mr. Disney isn't likely to tell his audiences that he's being stubborn in negotiating with his workers. But perhaps by the time this review appears, the controversy will be settled. . . . After all . . . we want to love the creator of Mickey Mouse and not have the reluctant Disney be [an] oppressing plant owner."[49] Even the rather gentle assertion that Disney was "stubborn" is startling in the context of the usual discourse about the Disney studio. The *Commonweal* review stands out as an interesting bit of textual analysis, finding meaning in what a film elided—the strike—rather than in what it chose to depict.

Nevertheless, *Commonweal* defused the power of the Disney strike to unsettle things, to turn workers against each other, and to raise issues about the relations between labor and capital. Former Disney animator, story writer, and director Jack Kinney has written that "Gunther Lessing, the company attorney, was hung in effigy. Cries of 'fink,' 'scab,' and other epithets were hurled against the nonstrikers, who retaliated by calling strikers 'commies.' It was a mess."[50] *Commonweal*, however, could only talk about the unrest as a breakdown in family systems and as a sign of personality flaws. Disney was clearly the punishing father/owner and the employees played the role of the children/workers—they were referred to as "his." The negotiating stalemate was the result only of Disney's unfortunate stubbornness. In bringing up the strike *Commonweal* appealed to the conscience of a readership concerned with social issues but most likely unwilling, because of its class privilege, to consider labor unrest a sign of a problem in capitalism itself rather than an example of a localized, temporary squabble.

The Corporate View

Rather than *Commonweal*'s "human person," *Business Week* concentrated on the corporation. The class of its clientele was hardly different from that of *Commonweal*, but the magazine's emphasis was decidedly more conservative. To attract advertisers, *Business Week* charted an executive profile of its audience, emphasizing "the corporate responsibility of its readers, their board memberships, their income and investments, and their influence on purchasing."[51] *Business Week* developed a kind of article that analyzed "the policies, problems, structure, and finances of a single corporation," and described itself as "preeminently the businessman's journal." With this self-image, and with the corporation from the businessman's point of view emerging as the main character in most of its articles,[52] *Business Week* could certainly mention strikes, but could hardly be expected to take a prolabor stand.

Business Week followed form in reporting work at the Disney studio and the strike that hit it. In March 1940, the magazine headlined "No Dust for Disney: New Studio City Features Elaborate Air-Conditioning Job to Cut Peculiar Production Hazards." The rest of the article detailed the extraordinary safety and comfort precautions taken by Disney in building a new studio for his 1,200 employees, with special emphasis placed on his state-of-the-art air-conditioning system. *Business Week* explained that the new technology had been developed "by engineers to cut down dust, dirt, noise, stray draughts, and other hazards to the peculiar type of movie production turned out by the creator of *Snow White* and *Pinocchio*."

Just from this information in the article's second paragraph, the reader might have suspected that the new system had been installed out of concern for the workers, while wondering precisely what special hazards animators may have faced. Immediately after the description of the system, however, the magazine explained that it had been installed not so much for the workers as for the product: "Disney photographs a sequence of scenes drawn and colored on celluloid . . . and a single particle of dust on a painting produces light effects and halations which spoil the picture." Because of this risk, part of the new system included a "de-dusting chamber where employees and visitors are frisked for dust." "Camera rooms will be cleaned and polished to the nth degree," the article claimed, and "floors will be waxed and employees wear lintless clothing."[53] These procedures, too, were taken for the sake of Mickey and Donald. For *Business Week*, the Disney studio became a fabulous place to work not because of concern for the employee but because of concern for the product.

The first mention of any labor unrest at the studio came a little more than a year later, in June 1941.[54] A photograph showed striking cartoonists carrying placards ranging from the purely informative—"Disney Studio On Strike"—to the more creative—"1 Genius against 1200 Guinea Pigs" (fig. 16). The article accompanying the photograph was really only a caption. It was all in italics, just like the descriptive captions under the photographs of the new equipment in the article about air-conditioning. The caption began by stating that "Mickey Mouse and Donald Duck were still limping along at less than full production this week," with the mention of the cartoon stars at least gently condemning the strike for bringing discomfort to lovable cultural icons. More decisively, the caption reaffirmed *Business Week*'s primary concern with product over the people making that product. Here, the striking workers were discussed primarily in relation to their impact upon Disney's film schedule.

The rest of the description deflected attention from working conditions and emphasized the squabbling between "the A.F.L. union's claim" on the animators' membership and that of "the American Society of Screen Cartoonists." In looking for acceptable ways to tell its readers about the strike, therefore, *Business Week* concentrated on work rather than workers and on union versus union instead of labor versus management, and reduced it all to caption-size.

By 1945, *Business Week* had diminished the strike to a virtual footnote in a studio profile called "But Is It Art?" That article extolled the studio, likening it to a big family with Disney as the corporate boss/benevolent

Figure 16. "Mickey Mouse and Donald Duck were still limping along at less than full production this week—while the A.F.L. Screen Cartoonists Guild picket the studios of Walt Disney Productions" (*Business Week*, 14 June 1941).

father. Readers learned that the 1,700 preferred stockholders in the studio and its enterprises were all "outsiders," that the "controlling common stock-holders . . . all are named Disney," and that "Disney-dominance in corporate affairs continues unbroken." Here the article reassuringly implied that the studio was something of a latter-day mom-and-pop store, with all non-Disneys reduced to secondary power positions as "outsiders." Indeed, the article went on to stress the "family control of the common stock," and then likened Disney to the sacrificing father who uses all of his paycheck to guarantee family welfare. That control, the article said, "has enabled Walter Elias Disney to perpetuate the plowing back policy [putting all earnings back into the corporation] he has followed during most of the 25 years since his animal caricatures sprang to life on the screen and parlayed his modest investment in cash and confidence into a seven-million dollar business." In this view of Disney history, the patriarch found himself rewarded with huge profits because he never sought to declare a profit at all. Emphasizing the point, the next paragraph explained that "not since the various Disney enterprises were consolidated in 1938 have Walt, his brother Roy, and their wives declared a dividend on the 355,000 outstanding shares of common which they hold exclusively."[55] Thus, the corporation equalled the family, and family interests led to corporate health (fig. 17).

Mention of the strike came two full columns after this homage, *Business Week*-style, to the American family. Under the heading "Show Goes On," the article said that "despite union troubles a couple of years ago and man-power shortages, which have seen the payroll at Disney productions shrink from a peak of 1,200 in 1940—when the enterprise moved into its new, air-conditioned, dustproof home—to something less than 900 today, there has been no disposition to curtail the entertainment output of the Burbank studios."[56] The strike was dismissed merely as "union troubles" and the "man-power shortages" were completely decontextualized. A 1945 reader might have assumed that the shortages came from military enlistment, when they had resulted at least in part from Disney's policy of firing strikers once the "troubles" had ended.[57] Further, the strike was worth mentioning only to show Disney's perseverance, as he had managed to maintain production "despite" the unrest. In the same breath as the strike, the article celebrated those ultra-modern conveniences that had been discussed in detail four years before, with this juxtaposition of "union troubles" and an air-conditioned, new, and dust-free workplace implying that the workers had no grounds on which to gripe.

In a parenthetical phrase, and without historicizing Disney's relation with the unions, the article then assured readers that the studio had entered the modern age of labor relations. "Disney operates a closed shop," the article said, "and now deals with some 30 labor unions."[58] There was no indication, though, that these were developments that came about precisely because of the strike, and that Disney never fully reconciled himself to them.[59]

Making Labor Disappear

Some of the high-class magazines subscribed to a Disney ethos of whistling while you work and refused to mention the strike. *The Saturday Review*, for instance, catered to a small clientele similar to that of *Commonweal* or *Business Week*.[60] But the *Review*'s stake in labor relations was somewhat different. Founded as a literary magazine in 1924, the *Review*

Figure 17. "Walt Disney surveys the campus-like lawns that hold down dust at the luxurious studios in Burbank, Calif., paid for by the famous movie mouse and duck" (*Business Week*, 10 February 1945).

stressed more purely aesthetic issues over political ones and the magazine did not budge on this art for art's sake stance until the early 1940s, when Norman Cousins became editor.[61] Only then did the magazine routinely include articles on politics, business, and mass communications. During the period under discussion, however, which marks the last pre-Cousins years and the first few years of his editorship, the change in editorial policy was not yet complete. As a result, the *Review* may have been more willing to cover wider aspects of culture, but not necessarily in very different contexts from before.

There may be at least one other influence on the magazine's attitude toward labor. For its first eight years of publication the *Review* was, to varying degrees, under the control of one of the country's largest media corporations, Time, Inc. Henry Luce's company bought a significant amount of stock in the new magazine and "its editors were invited to share *Time*'s offices; and Roy E. Larsen, *Time*'s circulation manager, conducted its circulation and promotional campaign."[62] Even during the 1940s, several years after Luce's control had ended, it is doubtful that any magazine that had been so closely and so recently affiliated with a corporation as powerful as Luce's could take a prolabor view of any strike.

In 1942, when he reviewed Robert Feild's *The Art of Walt Disney*, a critic as astute as Jay Leyda seemed completely unaware that a strike had ever taken place, and could assert that "at the Disney Studios a group of imaginative, disciplined artists give their work an amount of time and concentration that make them the envy of the motion picture industry."[63] If Disney was the envy of the industry, it was precisely because he had resisted attempts at unionization for so long, as many Hollywood movie workers had negotiated union contracts by the end of the 1930s. But one of the points of Leyda's article, of course, was that the people at Disney's studio were not workers at all. Instead, they were "disciplined artists" who spent lengthy hours on the job not because Disney demanded it of them but out of their commitment to their art.

When these artists experienced problems, they were more aesthetic than practical ones. Leyda wrote of "the delicate balance between the creating artist and the huge mechanism of the Studio," but not to expose any imbalance between employee and employer. He meant to stress the difficulty of expressing artistic concerns in films that had to be produced with almost frightening regularity and that could not greatly differ one from the other. These artists might be forced, by the demands of the corporation, to turn out cookie-cutter cartoons. Indeed, this was a problem faced not only by those artists who worked for Disney but by the corporate boss himself. Leyda asked whether "the unceasing struggle of the whole organization," and this clearly included Disney, "to provide vivid outlets for unbound imagination [might] bind itself unavoidably to a single precise style or a single habitual attitude?" For Leyda, the workers and the boss struggled together against a corporate structure that threatened their artistry.

When Walter Wanger referred to the studio as a "league of nations" in the above-mentioned 1943 *Review* article, he made sure to point out that the Disney plant functioned as a metaphor for international harmony and also as a concrete example of domestic tranquility. Like Leyda, and in the face of all evidence, he claimed that the studio was a place where everyone worked together happily in an atmosphere always informed by the pressures and anxieties of war, but also, always, removed from them. While making training films for the armed services or prowar films for domestic audiences, the studio nevertheless functioned as a "fairyland plant" where "thousands are creating and delivering in the interest of the new education of the free world." Working in a utopia, the studio employees—including, implicitly, Disney himself—were bound together by shared assumptions about and aspirations for the Allied countries. Although this may have been at least partially true, it seems just as obvious that they bonded as well in a way that excluded Disney, and, in fact, was directed against him.

A few months later, in "Mickey Icarus, 1943,"[64] Wanger described the studio as a sort of Hollywood think-tank, with all energy directed toward the war effort. He wrote that "more experts, scientists, and technicians operate under Disney's roofs than in any other one organization in the universe." These workers functioned much like Leyda's artists. They could stand for everyone at the studio even though they made up a very small percentage of the actual workforce. In Wanger's article, the studio became a place where only the elite worked, a group of people doubly removed from day-to-day issues concerning labor conditions. As "experts" or "scientists" they clearly were not from the working class, and like the employees Wanger mentioned in his earlier article their concern was not so much with their own conditions as with the condition of the world. "Out of this amazing studio," Wanger said, "come films on meteorology, on countless forms of technical instruction, on the prevention of malaria, in fact on every subject that might be, and is, useful in time of war." Indeed, precisely this kind of workplace made the United States worth defending, as Wanger insisted that "if every American could visit the studio, he would have a new admiration for his country."

These two articles were the only ones that Wanger wrote about Disney for the *Review* during the 1940s. As a producer himself, Wanger may have had a stake in mystifying motion picture production for the public, in making most movie labor disappear and stressing, instead, inspired artistic and scientific creation. But it was also a sign of *The Saturday Review*'s political position that a producer would even be chosen to write about the industry. Indeed, accompanying Wanger's review of *Victory Through Air Power* was a photograph from the film. Wanger probably did not choose the photo and he almost certainly did not write the caption for it. Instead, it served as an expression of the magazine's belief system. The caption reads, "As Walt Disney sees the shipyards." The photograph shows a curving, diagonal line of tankers that extends beyond the bottom and top frames of the photo. A supply train runs by them and at the lower left are smokestacks. Nowhere, however, are there any workers. There is simply technology and industry in geometric lines and perfect composition, as if they existed by themselves, with no need of workers to make them go (fig. 18).[65]

In these high-class journals in which workers tended to disappear, labor occasionally merited significant attention. A 1940 *Atlantic Monthly* article, even in its title, provided the typical characterization of Disney as an inspired creator who nonetheless labored long and hard: "Genius at Work." But the article also extolled the amount of work done at the studio by other people, by musicians, animators, stenographers, and story writers. All of the jobs and all of the immense numbers ("if it is to be a picture to run ten minutes on the screen, you know that you will ultimately have to show the audience 14,400 pictures. If it is a picture to run one hundred minutes, as a feature picture often does, you need only make 144,000 final drawings")[66] were gathered to celebrate a heroic division of labor rather than an inequitable assembly line system. When collective work was stressed and also when it disappeared, the high-class journals always worked to ignore any possibility of employee discontent, the kind that led to the bitter Disney strike.

Luce and Labor

Even more than *The Saturday Review* or other journals like it, *Time* magazine, the mass-market cornerstone of the Henry Luce media empire, spoke the corporate belief system about work. Unlike any of the high-class journals, *Time* had a vast readership—a circulation of around 1,690,000 in 1948[67]—and had become a primary news source for many Americans. During the period under discussion, and similar to most other mass-market magazines, the audience for *Time* consisted largely of a population with no high-school education and, at best, a moderate income.[68] But the news and features in the magazine were developed under the aegis of Luce, a multimillionaire with an Ivy League education. As a result, the reporting in *Time*, directed at a middle- and lower middle-class audience, reflected the ideology of the upper class.

Time's first mention during the 1940s of labor conditions at the studio removed them altogether from the realm of real work. In a review of *The New Spirit*, Disney's 1942 Treasury Department cartoon about the revised tax laws that made first-time taxpayers out of less affluent wage earners, readers learned that Donald Duck, "one of the world's most beloved cinema actors" and the star of the film, "earns less than $50 a

Figure 18. "As Walt Disney sees the shipyards." Publicity still from *Victory Through Air Power* (*The Saturday Review*, 4 September 1943).

week." *Time* made it clear that it disapproved of the salary: "That miserable retainer not only has to support himself in the extravagant style to which Hollywood is accustomed, but also has to feed, clothe, and house his three adopted nephews." These salary revelations came in the opening of the review, but even if the reader mistook the tone for a serious one, all blame for employee conditions was quickly removed from the employer. The review assured the reader that the "reputation" of Donald's boss, Disney, "is anything but a pinchpenny's." Disney's non-animated employees might have argued the point but, anyway, labor/management relations quickly lost importance in the article. Readers learned that Disney "was asked by U.S. Treasury Secretary Henry Morgenthau Jr. to make a picture reminding U.S. citizens that millions of them are expected to pay an income tax for the first time this year." Donald's salary, then, rather than pointing out the legitimate demands of a disgruntled workforce, instead stood for the necessity for low- and middle-income workers to pay their taxes despite the hardship. As a result, instead of being criticized for paying poor salaries, Disney-the-corporate-boss received praise for exhorting the poor to pay "taxes to beat the Axis."[69]

When it did discuss flesh-and-blood Disney employees, and despite its more middle- and working-class audience, *Time*, just like the high-class journals, concentrated on the privileged positions at the studio. With *Time* reaffirming Disney's position as benevolent father, a review of *Saludos Amigos* noted that "to make the film (and others to come) Disney took 15 of his staffmen on a three-month, 20,000-mile tour of South America." The life of the tourists seems to have been anything but difficult, as "they hobnobbed with artists and musicians."[70] The article turned Disney and his crew into happy, creative travelers, just the effect Disney hoped to achieve. Jack Kinney, who remained at the Burbank studio to work on *Dumbo*, has written that "during the strike negotiations, the studio decided that it would be best to get Walt out of town . . . [so] the State Department . . . encouraged Walt to go to South America, which he did, along with a handful of studio people."[71] Disney's trip obscured the labor unrest at the studio while presenting an image, eagerly transmitted by the popular journals, of absolute labor contentment.

In "Mickey's Coworkers," from 1945, *Time* proved that Disney was indeed no "pinchpenny." Discussing Disney's employment of Nelson Eddy, Aldous Huxley, and other high-art and popular culture superstars, *Time* said that the "new top-drawer talent will be paid on a picture-to-picture basis," and added, tongue in cheek, "probable pittance for the literary help: $2,500 a week and up."[72]

Time mentioned the animation unions just once during the decade, and then only to blame them for spiraling cartoon costs. In a 1946 article called "Stuffed Duck" appearing in the "Show Business" column and sandwiched between reports on "Foreign Trade" and "The Economy," the magazine said that "from the swank Disney Studios . . . came a new wave of raucous sound effects. . . . Most of the noise came from representatives of 27 unions protesting last week's layoff of 450 Disney employees, almost half the studio's staff." This was the "man-power shortage" that *Business Week* referred to so obliquely, but here *Time* clearly related it to union unrest. That unrest, though, simply became "noise," and rather than seeking explanations from union leaders the article referred to a management representative, "General Manager John F. Reeder." He said that "the new pay schedule . . . put into effect on the demand of the Screen Cartoonists Guild would not allow the studio to keep on going full blast with a reasonable hope of profit." *Time* added that the raise came to "a 25% increase—an estimated $1 million-a-year boost in the payroll." Thus, because it complied with the unions, the studio found itself on the verge of ruin.

Reeder continued the doomsday scenario, insisting that "work would have to stop . . . on all but four feature productions. . . . Workers on all other projects would have to go." The implication is clear: in forming a union the workers had cut their own throats and were practically forcing the studio to fire them. Rather than providing the union point of view, the article then placed Reeder's assertions in a context of cartoon economics in general while continuing to place the blame on studio employees. "Since 1940," *Time* said, "cartoon costs have jumped 165% (due almost wholly to increased labor costs) while revenues have increased only 12%. . . . Full-length cartoons, for all the fanfare about them, have only dug the hole deeper." Those labor costs, of course, came as the direct result of increased union activity and organization. In one of its only references to Disney's peers, *Time* then asserted that all cartoon bosses had been victimized by their workers. In *Time* and other journals Disney may have stood alone as a cartoon artist, but to underscore the dire problems felt by the heads of the cartoon industry the article said that "most cartoonmakers (e.g., [Walter] Lantz, [George] Pal, [Fred] Quimby, [Edward] Selzer) are worrying how long they can go on."[73]

While the bosses worried the employees enjoyed the pleasures of Disney's studio. For the photograph accompanying the article, the caption said "Disney Animators at Work." Their activity hardly seems like labor, however. The photo shows nine men, eight of them seated comfortably and all of them drawing, surrounding a nude female model whose arms are placed strategically but who poses provocatively nonetheless. Even though they had caused the fire, these employees doodled while the studio burned (fig. 19).

Labor as Labor

Unlike *Time* with its depiction of ungrateful artists, *Popular Science* fully recognized that making cartoons was all work. In this mass-market magazine founded on the idea of "diffusing scientific knowledge" to the people, and which by the end of World War II had a circulation of over one million, work became heroic, powering the technology and industry which, particularly during the war, formed the main focus of the magazine's coverage.[74]

Most of the magazine's articles about animation during the 1940s concentrated on how certain kinds of cartoons—training films, for in-

Figure 19. "Disney Animators at Work: Distributors have cold shoulders" (*Time* magazine, 12 August 1946).

stance, or live-action/animation combinations—were made. Each article stressed the worker's productive role but never any aspect of his/her relation to management. An article about *The Three Caballeros* called, with typical *Popular Science* flair, "How Disney Combines Living Actors with His Cartoon Characters," talked about the committees of people who made the film. When "layouts, after many conferences, have been approved, the background artists get the 'go' signal," the article explained. Then, "at the same time, the animators get busy. They work in front of mirrors and make faces at themselves, seeking to capture in their pictures the expressions that they wish their characters to register."[75]

So far, unlike any of the other journals when they were reviewing a Disney cartoon, Disney himself had not been mentioned as part of the creative process. Instead, the magazine romanticized the committee nature of the work, turning it into an adventure full of "go signals" and people "getting busy" and grown men making faces in the mirror in order to do their jobs effectively. There was no effort to hide labor here, as there was when *Time* attempted to make animators look like comfortably seated men sketching nudes. However, neither was there an attempt to discuss the real conditions of labor—salary, for instance, or hours, or the full hierarchy of jobs from Disney on down. There was simply the joy and excitement of production.

In similar fashion the rest of the article detailed all of the stages involved in production. Readers learned that "an assistant cleans up the [animator's] rough drawings. . . . Later these are traced by trained girl artists onto transparent celluloid, after which scores of girls ink the characters in lines of various thicknesses and colors." Along with this description of jobs and the indication that women were among the most exploited and least highly regarded of animation workers, the article provided numbers: "1000 compounded color formulas . . . 150 gallons of paint . . . $5 a pint."[76] Then, from a caption for one of the photos accompanying the article, readers learned that after Panchito, a main character in *The Three Caballeros*, had been "okayed . . . hundreds of studio workers began turning out the more than 150,000 paintings that were needed."[77] Here, as in most articles in *Popular Science*, the operative term was worker rather than artist. Disney's division of labor, however, and also his assembly-line methods became heroic rather than alienating and led to a remarkable technological artifact—a feature-length film that combined live action with animation—rather than to worker unrest.

In its last paragraph, and after this exhaustive cataloging of tasks, the article reminded its readers that the cartoon finally would be marked by the disappearance of these workers just as surely as they vanished from the shipyards in that photograph in *The Saturday Review* from *Victory Through Air Power*. *Popular Science* asserted that "one of the more outstanding qualities of the Disney pictures" was their "apparently effortless production." Then, even in this magazine that marvelled at anonymous, collective production within the corporate structure, the corporate boss finally became the central figure, the one who would be the most creative. "Will mystification outweigh story interest," the article asked in its conclusion, "or will Disney's genius make plausible the mingling of animated pictures with equally lively people?"[78] The film had become, simply, a product only of Disney's "genius," in complete contradiction to everything that the article had said before, but completely in keeping with a process of "mystification" carried out in most of the poular journals during the period.

We have come practically full circle. A high-class magazine such as *Commonweal* could mention the strike at Disney but only as a family crisis. *Business Week*, from the same category, hinted around the strike while concentrating on the necessity, even the marvel, of constant production. *Popular Science*, at the other end of the popular journal class scale, celebrated assembly line production but finally, like the other journals, focused on Disney as the sole creator, as the corporate *pater familias*.

Creating Disney

I cannot claim that the public read any of these popular journals uncritically and simply accepted all of the Disney data as true. Furthermore, my study has not dealt with those journals that themselves may have disputed the findings of *Time* and the others; that is, the journals of the left-wing press, such as *The New Masses*. Whatever the response of the readers, however, and whatever the opposition from less mainstream journals, the hegemony of the popular press is remarkable. Regardless of class affiliations or special interests, the journals from the period constructed animation and Disney almost identically.

At least in part, this indicates a journalism industry that, while catering to a vast and varied audience, was controlled by a select group of corporate magnates. Diversity in the case of the journals served as a mask for monopoly practice. But Disney discourse in the journals also pointed out Disney's unique status in American culture, a status that the journals both reflected and created. The sheer number of articles about Disney during the period—certainly there are more about him than about any other film producer—demonstrates the fascination that the public had for him.

It became the job of the journals to feed this fascination, to give the magazine-buying public what it wanted. It also became the journals' job, however, to control that fascination. This is not to say that Disney necessarily signified anything potentially subversive. But he did signify a great many things, because of all of the products manufactured under his name and because of his connections to art and commerce and also to politics and escapism. As a result, and under the guise of aesthetic judgment or business profile, the journal articles took part in an effort to lock in the possible readings of Disney. They made him comprehensible to the public in ways that conformed to a belief in benevolent business, racial tolerance, and strict class divisions at home coupled with a faith in United States political and commercial superiority in the global sphere.

5 | DISNEY DIPLOMACY

The Links between Culture, Commerce, and Government Policy

Popular and academic film histories recently have rediscovered the motion picture studio. Thanks to Thomas Schatz, Douglas Gomery, Tino Balio, and others, we have gone beyond the notion of the movie studio as oppressor of the individual artist.[1] Instead, we now have an excellent understanding of the day-to-day activities of MGM, Paramount, United Artists, and the other major and minor filmmaking companies. Still another kind of film studio operated during the classical period, however, one that worked in concert with all of the others: the United States government. And just as each of the commercial studios tended to be identified with certain styles and genres, so too was this federally funded one. The government-as-studio specialized in two modes of propaganda production, one foreign and the other domestic.

Walt Disney emerges as a case study for interpreting the intersection of cultural production, commerce, and government policy. My data come from the Departments of State and the Treasury, which collected files on Disney primarily just before, during, and after World War II, and from the FBI, which kept tabs on Disney from the late 1930s until his death in 1966.[2] Reading through these government documents does not provide us with the hidden history of Walt Disney. It is difficult to determine, for instance, whether some of the proposed events in State Department memos ever took place. Instead, the files reveal the federal government's efforts at social control at home and abroad and also its determination to maintain the global preeminence of United States industry.

In *How to Read Donald Duck*, their groundbreaking analysis of Disney's product in the world market, Ariel Dorfman and Armand Mattelart charted the impact of transnational capitalism on South America primarily during the 1950s and 1960s. But Disney first had become extensively involved in South America years before that. He began working for the State Department in 1941, during an era of shifting global alliances brought on by the war in Europe. At the same time, the closing of European markets to the Hollywood studios created a potential economic crisis for an industry that depended on foreign revenues. In response to this emergency, Disney went on a United States–sponsored tour of Latin America to assert the government's Good Neighborliness—that is, imperialism masked as benevolence—and also to make the region safe for Hollywood film companies. State Department documents, therefore, to use Andre Gunder Frank's terms, demonstrate the efforts of the metropolis—the United States—to maintain control of its satellites both as political allies and as economic dependents.[3]

Documents from the Treasury Department show the government endeavoring to create wartime consensus within the United States. My reading of these documents, however, reveals, despite the best efforts of the government, a wartime culture marked by class conflict rather than by unanimity. The Treasury's attempts to exercise social control through Disney animation serve as an example of increased government willingness during the 1940s to propagandize the domestic population. Indeed, the FBI documents about Disney prove that this willingness came about directly as a result of laws enacted under Franklin Roosevelt during the 1930s and 1940s that gave government institutions greater powers than ever before to monitor the governed.

The Disney documents articulate the responses of the government to a series of crises in United States–style capitalism: shifts in global markets due to the war and a collapse of confidence in the home economy during the Depression, a collapse which arguably made the government more determined than ever before to guard against real and imagined threats from the left. Further, like the new histories of the Hollywood studios, the files on Disney reveal that we cannot talk simply about art and popular culture divorced from business concerns. Cultural production, business, and the government are inextricably connected, and, in fact, it is frequently the common interests of business and government that produce popular culture.

The Big Bad Wolf as Good Neighbor

We know that D. W. Griffith, in 1915, used President Woodrow Wilson's enthusiastic response to *The Birth of a Nation* ("like writing history with lightning") to publicize his Civil War epic. A few years later, the motion picture industry chose Postmaster General (and confidant to President Warren Harding) Will Hays as the first president of the Motion Picture Producers and Distributors Association of America, in order to lessen the possibility of direct government intervention in the filmmaking business and to give the industry a sort of governmental imprimatur in the mind of the public. More significant, in terms of my interest here, the government itself, by the time of Disney's work for the State Department, had for many years been engaged in making films. During the Depression, for instance, the Roosevelt administration helped the major studios produce short subjects in support of the National Recovery Administration. So when the State Department decided to make films for South America, and to hire Disney for the job, it was simply following precedent rather than embarking on a boldly new project.[4]

All of the major movie studios, and not just Disney, were eager to cooperate with government plans to control Latin American film markets. At least from World War I, U.S. films dominated in the region; by 1935, Hollywood films accounted for more than three-quarters of the feature films exhibited in Argentina, and for 80 percent of the features shown in Mexico.[5] By the time of Disney's tour, however, the studios had become obsessively interested in South America because of the precariousness of European markets; Germany and Italy no longer imported U.S. films, and it was unclear whether Hollywood studios would have free access to occupied France or whether Great Britain, under siege from Germany, would remain open to Hollywood products.

At least in this early phase of the State Department's film production efforts, Disney, among all the Hollywood executives, played the most significant role in United States propaganda production in South America. The State Department had established an Office of Cultural Relations and appointed Nelson Rockefeller the coordinator of Commercial and Cultural Relations between the American Republics. Reporting to Rockefeller were representatives of related organizations, such as the Motion Picture Society for the Americas. All of these bureaus and offices and societies were designed to assert the presence of all manner of United States motion picture products— features, shorts, documentaries, etc.—in South America, and to make films about the region for U.S. distribution. Disney began to figure specifically in Rockefeller's plans in 1941, but even before that various State Department representatives were discussing Disney-type films. A Memorandum of Conversation from May 7, 1941, recording a discussion among several of these representatives, detailed the formation of government aesthetic norms for films to be produced in the United States for viewing in South America. Hollywood producer Kenneth Macgowan, who represented Rockefeller's Office of the Coordinator of Inter-American Affairs, said that "it was not his idea that the United States should seek to emulate the Nazis in their showing of films like *The Power of Thought*." Macgowan felt that "we should not display 'terror' films, designed to intimidate," and Pierre Boal, another participant in the conversation, agreed, suggesting films like "*Abraham Lincoln*, [and] *Edison, the Man* . . . which if shown convey a favorable impression of this country." The area had already seen too much of "preparedness, military expansion and naval expansion in the news reels and on regular amusement programs," according to Boal, so he stressed the importance of films of a different kind of "propagandistic character."

Seven months, then, before the United States entered the war, the lines of battle were already forming. Largely through the distribution of such films as *The Power of Thought*, Germany had created a strong propaganda presence in South America, and the United States felt the need to go into propaganda production itself in that area. With their emphasis on Lincoln and Edison, the participants in the conversation proposed a Great Man approach to films for South America in order to differentiate United States propaganda from the German variety. In doing so, Boal, Macgowan, and the others prepared for the government's use of Disney, who became the Great Man of U.S. culture when he toured the region.

This memorandum is one of the few State Department documents to mention German film efforts in South America, and in attempting to establish the United States as a filmmaking competitor with Germany it hints at the struggle for control of the area. For the State Department, Argentina and Brazil had become major battlegrounds, and the government ultimately made these two countries the major stops on Disney's tour. In Argentina, by the early 1940s, England's traditionally strong hold over the economy was beginning to weaken at a time when, because of the war, Great Britain was most in need of the foodstuffs it imported from that country. At the same time, because of a position of profascist neutrality, Argentina became a natural target of German wartime expansion. With English control over Argentina becoming more fragile, and with a new and threatening German position there, the United States made a series of mostly failed attempts during the 1940s to bring Argentina into its own sphere of influence, both commercially and through defense agreements.[6] Putting the emphasis on commerce and pleasure rather than on defense, however, Macgowan went on to stress in the memorandum the importance of "good short [films] in color" about technology, about making nature work for you, and about leisure activities; Weyerhauser Company's *Trees and Houses*, for instance, could be "made acceptable by editing," and Macgowan also brought up "the film he desires to make on the American pleasure park . . . giving due care to avoid anything which might indeed cause laughter *at* us, instead of *with* us."

In establishing a propaganda aesthetic for South America, Macgowan was discussing Disney films before Disney had come into the picture. Disney had been one of the pioneers in the use of Technicolor, the short film was his stock-in-trade, and, of course, his films generally were comedies, the kind designed to provoke laughter "with us." This early memo also hinted at a major problem in the films that came out of Disney's tour of the area, and one that characterized U.S. policy in South America in general, that is, an only casual sensitivity to issues of cultural difference.

At the end of the conversation, the memo indicated that "an amusing and illuminating illustration rose of the difficulties of soundtracking in Spanish for display through all the . . . republics where Spanish is the basic language." A suggested Spanish title for a film would pass muster in most countires, but would constitute an insult in Mexico. In all of the documents, this is the only mention of differences between Latin American countries, and it implies, as well, that the United States understood the linguistic complexity of the area and would be able to make films for all of the countries in it with just a little tinkering with individual versions.

A few months later, in July 1941, Disney had begun to figure in government plans to depict the culture of its Good Neighbor. John Hay Whitney, a millionaire who had dabbled in movies for years (he fronted much of the money for David O. Selznick's independent movie studio in the 1930s), outlined the objectives and beliefs of the Office of Inter-American Affairs in a memo to the undersecretary of state. Whitney wrote that, as part of a project to produce films "portraying Latin American life for distribution in the United States," "Walt Disney will produce a series of twelve one-reel subjects on Latin American countries. These pictures will be started, and produced in part, in the Argentine and will be released over a period of two years. . . . At the present time, Disney is also considering a feature-length picture of great interest to Latin America, namely *Don Quixote*."

For Whitney and Disney, Latin America was simply an extension of Europe. To them, it made perfect sense for a narrative from Spain—*Don Quixote*—to be filmed in South America, and they assumed, as well, that such a film would be "of great interest" to the Latin American audience. Spain's former status as an imperial power, which created the conditions under which many in that audience came to speak Spanish, apparently made no difference to Whitney or Disney. Instead, this film project (which in fact never got much beyond the planning stage) indicates that, for Rockefeller's commission, Latin American culture was Spanish culture, and European cultural artifacts, such as the work of Cervantes, could be exported without complications to non-European countries and could represent those countries' traditions and histories.

Whitney justified Disney's project and others like it by assuring the undersecretary that "we have had a government official of Peru ask us to make films about Peru and show them in Peru." Thus, for Whitney, a government official could represent the desire of an entire country, with Whitney then able to conclude that Latin Americans indeed wanted the United States to make films about them.

Explaining the use of Disney's films in the United States, Whitney insisted that "when we produce . . . motion pictures about Latin America and distribute them to our people . . . we give Latin Americans tangible and flattering evidence of our interest and respect." Whitney, of course, was talking here largely of cartoons, hardly privileged cultural artifacts. Nevertheless, he was repeating the flawless calculus of cultural imperialism. Walt Disney, a representative of the United States, could tour a foreign culture, come to understand it in just a short time, film it, and then bring it back home with him, all with the blessing and thanks of the culture he had visited.

Disney developed two feature-length films from his trip South, *Saludos Amigos* and *The Three Caballeros*, and several educational short films (with such titles as *Hookworm*, *Cleanliness Brings Health*, and *How Disease Travels*). The State Department apparently believed that these films accurately depicted life in Latin America, despite the criticism of one of the men who worked for Disney during this time. Florencio Molina Campos, an Argentine artist whom Disney hired and brought to Los Angeles, wrote a letter that, through various channels, finally made its way to Secretary of State Cordell Hull. Molina Campos sounds exasperated, saying of his work at the studio, "what will come of it all, not even God knows!" He then complained that, "one picture in particular—I think it will be called *Goofy Gaucho*—was brought to my office so that I might study it and arrange the gaucho dialect that was to accompany the version destined for Spanish-speaking America. I tried to find a way of fixing it, but I found all my efforts so hopeless that I told them I didn't see any way at all; such was the conglomeration of errors." The artist found that he was little more than public relations window-dressing at the studio, because by the time he was hired to provide authenticity to the films, "the larger part of the work had been finished." When he complained to Disney, the producer "had no real answer, making a vague gesture like a child caught in some prank."

The film that Campos discussed finally became the "El Gaucho Goofy" section of *Saludos Amigos*, a film divided into four separate cartoons. In turning Goofy into a gaucho in order to characterize Argentine culture, Disney was making free use of a national icon whose importance he clearly did not understand. Gauchos had functioned as a kind of rural proletariat in Argentina during the latter half of the nineteenth century, doing much of the country's manual labor. By the early part of the twentieth century, Argentina's population had been largely urbanized and so clung to the gaucho as a kind of Argentine essence in much the same way that people in the United States romanticized the cowboy. A major Argentine literary historian of the time, Ricardo Rojás, claimed that "genuine Argentine thought had been born" in the gaucho, and the novelist Ricardo Guiraldes, who in the 1920s became one of Argentina's most famous writers, extolled the gaucho as the "true Argentine."[7] Of course, the word of a literary historian and that of a novelist do not prove that the gaucho was an absolutely unproblematic icon in Argentina during the early 1940s. These men do articulate, however, an interest in defining a national identity that was very probably widespread in a country that was rapidly changing its own demographics from country to city, and, in addition, changing its relations with the rest of the world. But while Campos was striving for national character, Disney, the major film producer in the government's project to portray life in Latin America, was interested only in caricature.

It would be far too simple to claim that Disney, as the representative of U.S. interests, was simply a racist, while Campos stood for Argentine integrity. Furthermore, a letter written to Secretary of State Hull by F. G. Klock complicates Molina Campos's complaints. Klock's official position is unclear, but he apparently knew Molina Campos because he mentioned some of the details of the artist's working arrangement with Disney. Klock explained that "if we want to sell phoney gauchos to the Americans that is Disney's business, but I believe we should put a stop to the insulting of our neighbors and the revealing of ourselves as having little knowledge of their lives and habits, while we pose as good neighbors." He then insisted, however, that the real problem with "El Gaucho Goofy" is that it featured the star "playing a guitar which is gaily decorated with colored designs," and that "it so happens that gauchos are real he-men and under no circumstances would be seen with such a guitar, which immediately labels them as drug-store cowboys." This complaint may well have come from Campos, and the references to "he-men" and "drug-store cowboys" demonstrate that concerns with Argentine national identity were more than just a little tinged with rather dubious notions about the representation of masculinity.

More accurately, the personality clash between temperamental employee and childish employer exemplified the contradictions inherent to the State Department's project: Campos's version of authenticity versus Disney's concern that his film conform to cartoon character and narrative conventions in order to maximize profits; an office administered by such multimillionaire businessmen as Rockefeller and Whitney that is charged with representing an ethnic and cultural other; a perhaps benevolent Good Neighborliness and also an understandable desire to combat Nazi propaganda, tempered not only by racism but by the need to secure South American markets for Hollywood films and other United States exports; and the desire to stimulate South American commercial activity while assuring the dominance of United States economic interests.

Culture as Commerce

Rockefeller's office was extremely sensitive to this last issue. In his lengthy memo of July 26, 1941, Whitney explained a system of economic domination in terms of U.S. benevolence, with Disney as emissary. The goal of the Office of Inter-American Affairs was "to employ motion pictures as a medium for the development and improvement of the Commercial and Cultural Relations between the Latin American Republics and the United States." Further, "emphasis will be placed on the benefits, culturally and economically, to be derived by the Latin American countries through the improvement in our relations." Getting down to specific cases, Whitney planned to send Disney and a crew of fourteen to Latin America, where "they will visit Brazil and possibly Chile and set up shop in the Argentine where they will work for a period of from four to six weeks." Once established in the area, Disney would "endeavor to find outstanding national cartoonists to survey the possibilities of establishing a company in the Argentine." His principal project was to "produce a series of twelve one-reel subjects on Latin American countries. These pictures will be started, and produced in part, in the Argentine. . . . The pictures will be produced in color, hopefully in cooperation with outstanding Latin American caricaturists such as Campos (Argentina), Delano (Chile), and Covarrubias (Mexico). The musical background may be by such outstanding Latin American composers as Villa Lobos (Brazil) and Chaves (Mexico)."

Token hires, then, of one or two artists from each country apparently were intended to signify for Whitney economic benefits for South America, while satisfying the demand to find those outstanding national cartoonists. With Molina Campos's story we have already seen the use that Disney made of these cartoonists, indicating that, at least on his part, there was no interest in stimulating animation activity in Latin America by training local artists in cartooning. Whitney asserted that "Walt Disney's project . . . will give impetus to production in South America," but it becomes clear that rather than encouraging the creation of an independent film industry in the area, Disney's tour was designed to test the prospects for extending U.S. domination of the cinema there. Movies made in Hollywood had monopolized the region's screens for many years; Disney's tour, however, marked a tentative effort to control film production *within* Latin America. After discussing Disney, Whitney made a brief mention of "short subject productions in Latin America by M.G.M., Pathé and Paramount," and added that "Warner's has agreed to undertake similar work in the near future." Improving the region's culture and commerce, therefore, meant allowing U.S. film companies to begin production there and to provide United States–made entertainment.

To make sure that Hollywood-style production could take place under U.S. corporate auspices, Whitney added that "the president of Technicolor, Inc. has indicated that he may survey Latin America and . . . establish a Technicolor laboratory in one of the countries." Rather than help to guarantee U.S. domination not only of film production but also the sale of raw film stock, this move by Technicolor, according to Whitney, "would aid our efforts because South American beauty can best be shown through color photography." Whitney then underscored the absolute centrality of Disney's tour to this larger project of motivating other studios and film-related companies such as Technicolor to establish themselves in South America. Discussing star tours as a means of public relations, Whitney mentioned that Charles Boyer, Claudette Colbert, and perhaps Olivia de Havilland would be visiting Mexico for the premiere of the film *Hold Back the Dawn.* "Originally the tour was planned for August," Whitney explained, "but in view of the Disney trip probably should be postponed." Disney's tour of the region could have no competition, even from an entourage sent by Rockefeller's office and engaged in the same project.

Disney's efforts can best be understood in the context of U.S. economic necessity and South America's entrance into modern capitalism. As I mentioned above, the U.S. film industry had to concentrate on the region's markets because of the war in Europe, and the government aided the effort because of its desire to counteract German propaganda there. Along with these economic imperatives brought on by the war, however, there existed the issue of South American nationalism. During the 1930s, a number of South American countries pledged themselves to depend less on foreign capital and more on their own industry. In Brazil, a country of major importance to Rockefeller's commission and therefore a major stop on Disney's tour, one of the programs that developed from the 1930 revolution "advocated the creation of an independent national economy established on a strong industrial base" and emphasized that "industrialization was associated with national liberation." This movement linked an agrarian-export economy with dependency on the United States and Europe and insisted that liberation could come only from industrial development.[8]

In terms of this development, several Latin American countries not only weathered the Depression of the 1930s, but emerged from it with marked economic expansion. Argentina, another main stop on Disney's tour, managed to add 30,000 kilometers to a road system that, in 1932, had only 2,100. This same period also witnessed a massive public works buildup along with significant increases in tourism and in the manufacturing of such substances as rubber and petroleum, and of equipment such as automobiles, trucks, and buses.[9]

Brazil, Argentina, and other countries envisioned the result of this industrial development as the economic independence of the region. All of the improvements, however, in addition to an increasingly urban populace, actually made South America an ideal place for U.S. companies to take advantage of the new roads and factories to deliver goods to a newly affluent clientele. In his memos about Disney, then, Whitney may have talked about portraying the natural beauty of the region and of stimulating industry there. But his use of Disney, Technicolor, Inc., and the major motion picture studios really functioned as one of many attempts by the United States to assert its dominance in the area at a time when many countries there were insisting on their economic independence.

A memo from May 10, 1941, clearly stated Whitney's project. The document described all of the committees of Rockefeller's office and explained that the Committee on South American Film Facilities "has as its ultimate purpose the production in whole or in part of United States pictures in Latin America." Later, the memo depicted Herbert Kalmus, the president of Technicolor, as eager to conduct a "survey of South America as a possible new field for technicolor operations." Then, stressing long-range plans, the memo said that many of the office's projects, such as "investments in Latin American motion picture industries, the exchange of talent, the education of young men who may later be sent to South America, will require a continuous effort and may not result in an immediate accomplishment." In describing the organization of U.S. production facilities in South America, in addition to Kalmus's interest and the army of men to be sent over to the region, the memo described full-scale technological and commercial imperialism.

Labor at Home and Abroad

By establishing filmmaking units, Disney would be one of the major participants in this gradual takeover. The memo goes on to say that "Walt Disney has indicated that he may visit Argentina, Brazil, and possibly Chile and while there start the production of a South American subject or series of cartoons," and then adds that, during his tour, Disney "might also score *Bambi* and perhaps another picture now nearing completion in the United States for the South American markets." For Disney, South America could become an oasis, a location distant from the Hollywood unions that made postproduction projects like scoring so expensive, and also a base of operations from which he could exploit a new and expanding market.

There were complications with Disney's enterprise, and these, too, almost certainly had to do with his relationship with the unions. The memo says that "the financial problems involved in this project have been referred to the Coordinator's office," with the term "financial problems" functioning as a tactful metaphor for the strike over the right to unionize that was going on at the Disney studio at the time.

The memos show that the government was deeply concerned with the unrest at the studio. In one dated July 1, 1941, Whitney explained that "the Walt Disney project has been approved by the State Department," and that "contracts are now being drawn and will be executed immediately upon solution of the Disney labor problem." In another memo written nine days later, Whitney seems more irritated, as he complained that "the Walt Disney labor dispute has continued to hold up execution of our contract." Disney biographers and some of the people who worked at the studio have claimed that the government sent Disney to South America in order to get him away from the labor negotiations and to generate positive publicity for him.[10] The memos show that the government had an even more active role during the strike, supporting Disney financially or at least planning to support him. Whitney wrote that "we are proposing to execute the contract immediately on the understanding that he [Disney] cannot afford to continue his present outlay and risk and that he will have to abandon the project if we cannot support him temporarily."

Whitney's plan for assistance can be interpreted at least two ways. The government felt that Disney's State Department films were so important to the war effort that helping management in a labor dispute could be justified. Or, the perceived urgent need to control labor practices in South America, coupled with Rockefeller and Whitney's own pro-management sympathies, convinced the State Department to work against labor in at least one isolated case in the United States. Almost certainly, if Disney did receive government funding during the strike it enabled him not only to begin work on his State Department films but also to hold out that much longer against his striking workers.

The Great Man

Perhaps to justify this involvement with Disney and also to gain favorable publicity for the Disney project in South America and to create the feeling that Disney was welcomed by the people in the countries he toured, State Department officials worked hard to construct Disney's uniqueness. Joseph F. Burt, for instance, the American consul in Chile, sent a letter on November 4, 1941, to Secretary of State Hull about a special diploma awarded to Disney by Chile's Society for the Protection of Animals. He included a translation of a speech by the society's president, who said, in part, "I take pleasure in enclosing . . . [a] Diploma which the Institution confers on Mr. Walt Disney, who recently visited our country, for his brilliant work and ever increasing admiration towards the irrational beings, to which he gives life in celluloid."

It is nearly impossible to judge the importance of the society that honored Disney, but its official discourse was completely different from that of Molina Campos, the Argentine artist who complained about Disney's animals. Campos's letter apparently received no response from State Department officials. But department representatives clearly were pleased with the society's award, as notification of the honor was sent to Washington by Burt, then relayed to Disney by Assistant Secretary of State Breckinridge Long, and then mentioned again in a letter of thanks that Long wrote to Burt. Thus, the award served as the kind of Great Man publicity that State sought to create.

Just as in the speech by the society's president, in documents written by State Department officials Disney emerged as a one-of-a-kind talent. On April 14, 1943, Charles A. Thomson, the chief of the Division of Cultural Relations, wrote a letter about Disney and animation to Walter Thurston, the American minister in El Salvador. Thomson explained that "in cooperation with representatives of the Office of the Coordinator I have been giving consideration to a proposal for a campaign against illiteracy which would utilize the cartoon type of motion picture developed by Walt Disney." It is not even clear here that Disney would work on the project; instead, his name was simply synonymous with cartoons, and he was given credit for having developed them, just as D. W. Griffith often gets credit for fathering the modern motion picture.

Thomson then pointed out, however, that "in order to work out plans for such a cooperative movement, it has been suggested that a seminar of two or three weeks' duration . . . be held in the Disney Studio in Hollywood. The seminar would be made up of five or six representatives from the other American republics—persons who combine imagination and ideas with a degree of practical experience in education." Disney here became the man who could show others how to combat illiteracy. In doing so he would be introducing South Americans to the production methods of his studio. Once again, the line between United States Good Neighborliness and cultural and commercial imperialism became blurred. South American Disney trainees would teach people in their countries how to read, but in doing so they would introduce United States motion picture assembly-line methods. As a result, these countries would doubtless depend on Hollywood technology and would remain sympathetic markets for Hollywood films, because their own movies would conform to the norms established by Hollywood and passed on by Disney, the master teacher.

Animation and Empire

Thomson wrote his memo about two years after Disney first started working for the State Department by taking his 1941 tour of South America. Indeed, almost all of the memos about Disney were written in 1941, and the government's use of Disney in that year, along with its concern over relations between the U.S. film industry and Latin America, point up just how pivotal that year was. In 1940, Argentina, one of the primary stops on Disney's trip, enjoyed a brief return to democratic principles, and that year and the next the United States worked hard, albeit unsuccessfully, for a competitive foothold in Argentina against not only the Germans, but the British.[11] By 1942, however, Argentina's hostility to the United States increased, resulting in a considerable drop in U.S. exports, and the 1943 military coup ended all U.S. hope of establishing a sphere of influence in Argentina for the rest of the war.[12]

Even with this strong effort around the time of the tour, the State Department realized that Argentina's neutrality made it absolutely necessary to look elsewhere in South America, and so the United States showered Brazil with economic and military aid.[13] The United States instituted a policy of supplying the Brazilians with "goods that were surplus at home and could thus contribute to the utilization of excess capacity in the export industry."[14] With the Hollywood studios largely having recovered from, and adjusted to, the Depression, and enjoying significant profits during this period, movies certainly would have been counted among those goods that could be produced to excess at home. Further, as early as 1932, when Great Britain severely cut down its exports to Brazil, the United States had been looking for ways to exploit markets there. Once the war began, the United States sought even more enthusiastically to enhance its own position in Brazil and to make sure that Brazil would remain an ally, thereby countering Argentina's neutrality (a neutrality that, in fact, shaded more and more toward a sympathy for the German side).[15]

In making Argentina and Brazil two of the major stops on Disney's tour, the United States, in effect, wanted to have its cake and eat it, too; that is, curry favor in a hostile country—Argentina—and in a friendlier one—Brazil. In both of these countries, and also in others in South America, nationalist movements had created hostility toward the United States. Disney, a benign, entertaining, and educational presence, gave assurance that U.S. interests were benevolent rather than self-serving. The industrialization of these countries during the 1930s, which took place in order to make them less dependent on foreign industry, turned them instead into perfect targets for U.S. corporations that could make use of ready-made factories, new roads, and newly-urban populations needy for consumer goods. Reading the State Department documents reveals that such figures as Disney played a part in the United States' efforts to assert even more strongly than before its control of South America. Indeed, Disney's participation demonstrates the similar interests of the government and private industry, and their ability (despite such domestic squabbles as the Justice Department's antitrust action against the movie studios) to work together smoothly on an international scale.

The FBI: Surveillance, Cooperation, and Control

State Department documents reveal an apparently unproblematic relationship between Disney and the government. The government sided with Disney in his labor dispute, and in exchange for some propaganda movies from his studio allowed him to expand his filmmaking operation into South America. Another government bureau, however, enjoyed a much longer and less sympathetic relationship with Disney. During the thirty year period that FBI officials monitored the activities of the film producer, they never really questioned Disney's Americanism, and, in fact, exploited it by making him a special contact for their agents. But the Bureau did have some doubts about some of Disney's relationships with people and with groups, and argued with him over the representations of G-men in his films. Much has been made of the FBI's investigations of various so-called unfriendly Americans. But as the Disney file proves, the Bureau investigated friendly ones, too, and in ways that help explain issues of government surveillance, government use of celebrated citizens, and the government's sense (or, at least, the FBI's) of its own image.

The FBI became seriously interested in Disney in 1941 during an investigation of the strikers at his studio, when the Bureau sought to determine whether any employees attempted to extort funds from their boss. According to documents from July 1941, Disney denied that the FBI's suspect, whose name has been deleted from the file, "ever demanded or received any pay-offs from Disney or his organization." The Bureau then paraphrased Disney's assertions that "due to the curtailment of the showing of his pictures abroad [because of the war], it was necessary for him to cut down on his staff of employees at the studio," and that, "as a result . . . he laid off approximately nineteen men, some of whom had been in his employ less than one year." According to Disney, those nineteen men "went around to the various other employees at the studio and stated that approximately two hundred were to be laid off," and that "as a result of this 'whispering campaign,' a general strike was called." The document discussing this issue is heavily censored, but it seems as if the Bureau was interested primarily in the management position on the strike. An agent duly recorded Disney's belief that the strikers had been duped by a handful of workers, and that Disney's management style was dictated by a wartime economy that cut off many European markets rather than by his own stinginess, or, at least, by his distrust of labor and labor unions.

Disney the Subversive

Despite this 1941 defense of Disney-as-corporate boss, the FBI began collecting potentially damaging information about Disney in 1943. In October of that year, in a report labeled "League of American Writers" and "Internal Security," the FBI examined the "Writers Congress held October 1–3 at UCLA, with 1,500 writers in attendance under joint auspices Hollywood Writers Mobilization and UCLA" (fig. 20). The congress sounds harmless enough; among other things, "resolutions were passed advocating creation of Department of Arts and Letters by U.S. Government, a cultural and educational congress to meet in Central or South America in the near future, [and] the development of cultural relations between the United Nations." The report noted, however, that "the Congress drew charges from [the] California Legislative Fact-Finding Committee of Communist instigation and control."

Disney's name appears as one of the "guests of the evening." Because of his experience in South America just a few years before, it made sense

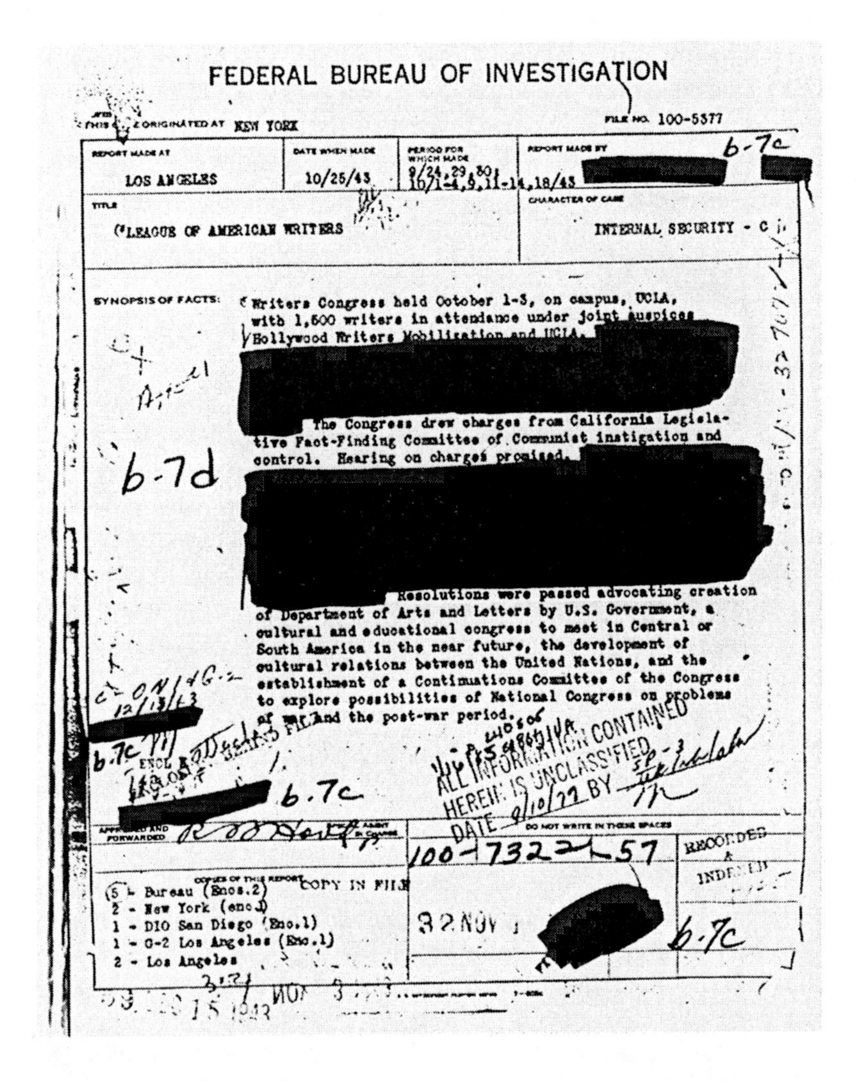
FEDERAL BUREAU OF INVESTIGATION

THIS CASE ORIGINATED AT NEW YORK — FILE NO. 100-5377

REPORT MADE AT	DATE WHEN MADE	PERIOD FOR WHICH MADE	REPORT MADE BY
LOS ANGELES	10/25/43	9/24,29,30; 10/1-4,9,11-14,18/43	b-7c

TITLE	CHARACTER OF CASE
"LEAGUE OF AMERICAN WRITERS	INTERNAL SECURITY - C

SYNOPSIS OF FACTS: Writers Congress held October 1-3, on campus, UCLA, with 1,500 writers in attendance under joint auspices Hollywood Writers Mobilization and UCLA.

The Congress drew charges from California Legislative Fact-Finding Committee of Communist instigation and control. Hearing on charges promised.

b-7d

Resolutions were passed advocating creation of Department of Arts and Letters by U.S. Government, a cultural and educational congress to meet in Central or South America in the near future, the development of cultural relations between the United Nations, and the establishment of a Continuations Committee of the Congress to explore possibilities of National Congress on problems of war and the post-war period.

b-7c

ALL INFORMATION CONTAINED HEREIN IS UNCLASSIFIED DATE 9/10/77 BY

b-7c

APPROVED AND FORWARDED — SPECIAL AGENT IN CHARGE

DO NOT WRITE IN THESE SPACES

100-7322-57 — RECORDED & INDEXED

COPIES OF THIS REPORT — COPY IN FILE

5 - Bureau (Encs.2)
2 - New York (enc.1)
1 - DIO San Diego (Enc.1)
1 - G-2 Los Angeles (Enc.1)
2 - Los Angeles

b-7c

NOV 15 1943

Figure 20. From Walt Disney's FBI file, a report on a 1943 Writers Congress, attended by Disney. "The Congress drew charges from [the] California Legislative Fact-Finding Committee of Communist instigation and control."

for him to attend a congress concerned with relations with that area. A look at the other guests shows not so much any single ideological line for the congress but rather the wide range of support that the writers enjoyed. Darryl F. Zanuck, Walter Wanger, and Jack Warner, Disney's fellow film producers, were there, but so were novelists Theodore Dreiser, Thomas Mann, and Lion Feuchtwanger, and radio personality Fred Allen.

A few months before, in June 1943, the FBI investigated the Council for Pan American Democracy, once again under the heading "Internal Security." The Council met for a "Night of the Americas" in New York, and the guest list is a surreal assortment of politicians, artists, writers, and businessmen. According to a flier in the FBI documents, the "Great Chilean Poet and Consul General to Mexico" Pablo Neruda was there, along with the actress Margo, Xavier Cugat, Paul Robeson, Orson Welles, Carmen Miranda, John Gunther, Norman Corwin, Disney, and various South American labor leaders, cabinet ministers, and elected officials (fig. 21).

A year later, in June 1944, and calling it a "Security Matter," the FBI filed a report on a tribute in New York to Art Young, who had just died, and whom the publicity for the event described as the "dean of American cartoonists." *New Masses*, a left-wing magazine, organized the tribute, and attendees included Paul Robeson and Howard Fast, both of whom held suspect political views, while the sponsors of the event were, among others, Ernest Hemingway and Disney. The FBI expressed concern that, along with Disney, "the hierarchy of the Communist Party will be present at this meeting" (fig. 22).

At this time, Disney was almost certainly not under any particular FBI surveillance, while Robeson and Fast probably were. But Disney's appearance in the FBI documents sheds some light on Bureau methods. For the FBI, surveillance meant targetting an individual and then keeping tabs on all of his/her activities. In the process of collecting this information, however, the FBI also gathered intelligence on everyone who came into contact with the surveillance suspect.[16]

Disney was just one of probably thousands of people who became the victims of this kind of scattershot intelligence gathering. A December 14, 1956, memo in Disney's file responded to a request for information regarding nominees to President Eisenhower's Citizens Advisory Committee on Fitness of American Youth. About Disney, the FBI explained that "no investigation has been conducted . . . concerning" him. This was true

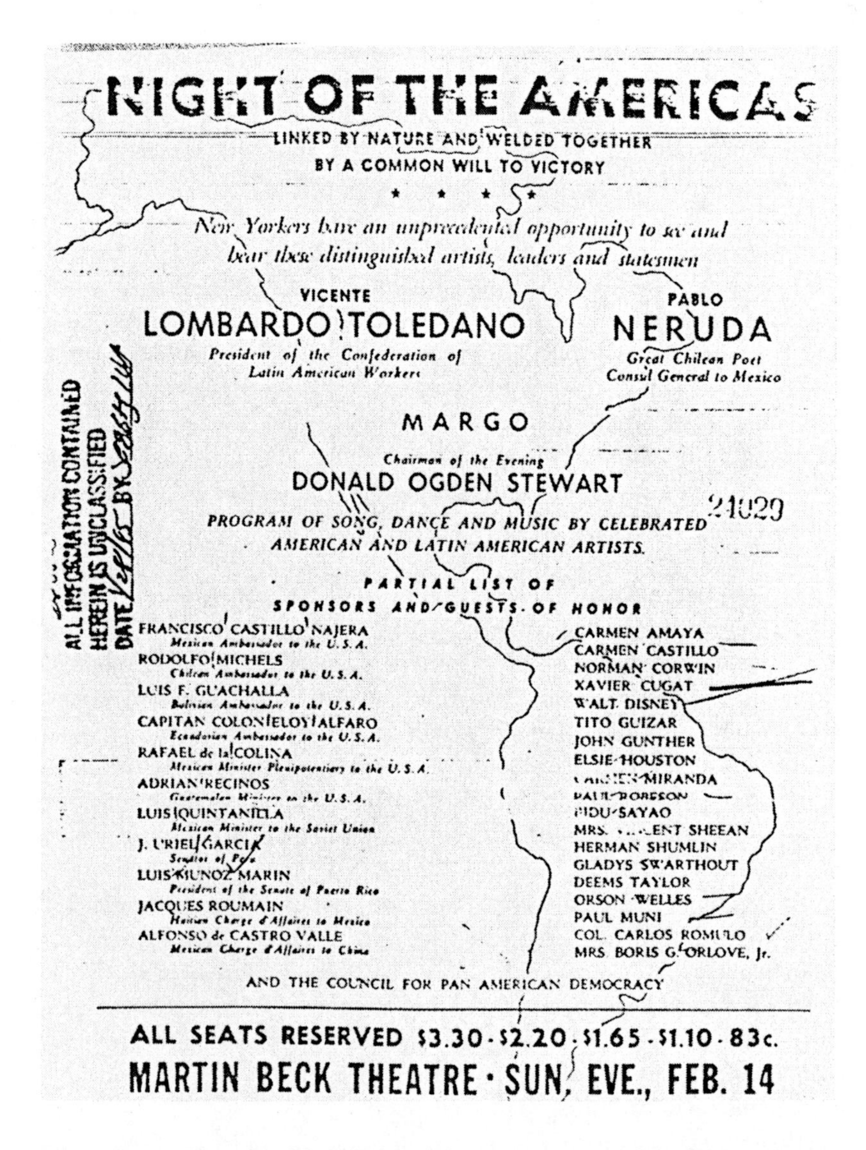

Figure 21. Flier for the "Night of the Americas," 1943, listing Disney as one of the "sponsors and guests of honor" (Walt Disney's FBI file).

FILE

b.7c

NEW MASSES
invites you to participate in
A TRIBUTE
— To —
THE MEMORY OF ART YOUNG
Dean of American Cartoonists
and contributing editor of New Masses
Sponsors Include:
Rockwell Kent, Carl Sandburg, Earl Browder, William Gropper, Hugo Gellert, Max Weber, Boardman Robinson, Paul Robeson, Langston Hughes, Mike Gold, Donald Ogden Stewart, Walt Disney, Crockett Johnson, Ernest Hemingway and Howard Fast.
PROMINENT SPEAKERS FROM THE
ART AND LITERARY WORLD
A Program of Dance and Music
MANHATTAN CENTER
Thursday Evening, January 27th, at 8:30
TICKETS 50 CENTS
Get Your Tickets Early at Workers Bookshop, 50 E. 13th St., Bookfair, 133 West 44th St., New Masses, 104 East 9th St.
There Will Be No Standing Room

246505
ALL INFORMATION CONTAINED
HEREIN IS UNCLASSIFIED
DATE 1/15/85 BY

This is a clipping from
page 14 of the
DAILY WORKER
Date Jan 13, 1944
Clipped at the Seat of
Government

61-9152-A
NOT RECORDED
85 JAN 17 1944

b.7c

27 JAN 19 1944

Figure 22. Announcement of the 1944 tribute to Art Young, which Disney attended. The FBI referred to *New Masses*, the tribute's sponsor, as "a nationally circulated weekly journal of the Communist Party" (Walt Disney's FBI file).

in the strictest sense, as Disney had never been the object of FBI surveillance. But the FBI cautioned that "this Bureau's files reflect the receipt of a flier issued by the Council for Pan-American Democracy advertising the 'Night of the Americas' to be held . . . on February 14, 1943. . . . The flier carried a partial list of sponsors and guests of honor which included the name of 'Walt Disney.'" The Bureau then explained the suspect status of the Council, and added that Disney also attended the tribute to Art Young, and warned that the periodical that sponsored the tribute, *New Masses*, was a "nationally circulated weekly journal of the Communist Party."

On July 26, 1951, the FBI filed a memorandum on Disney and sent the original to the CIA. It remains unclear whether the memo came in response to a CIA request or was simply sent along as something that might interest the intelligence agency. This memo, like the letter to the Citizens Advisory Committee, revealed that Disney attended the Pan American Democracy meeting and then went to the tribute to Art Young, which meant that Disney associated with suspect individuals. Similarly, in May of 1955, the Bureau sent a memo to the United States Information Agency, in response to USIA's request, providing assurance that Disney had never been the subject of an investigation but that he had attended both the Council for Pan American Democracy and the Young tribute (fig. 23). A 1964 memo repeated the same information, and while it cautioned that "this document contains neither recommendations nor conclusions of the FBI," it seems clear that the Bureau still believed Disney's attendance at those events to be of some significance. Finally, in a May 1966 response to Marvin Watson, a special assistant to the President, the FBI repeated yet again that, while no investigation had ever been carried out concerning Disney, he had, nonetheless, made some poor choices in how he spent his evenings out.

Spying at Home

Indiscriminately tracking and keeping files on prominent people such as Disney does not simply signify Bureau Director J. Edgar Hoover's unreasonable fear of Communism. This surveillance data derived from a five-year period during which the federal government curtailed individual rights and gave new power to the FBI. As late as 1936, Hoover could

MAY 10 1955
NAME CHECK

Name Checks

May 10, 1955

(WALTER E.) WALT DISNEY — Summary
Born: 1901
Chicago, Illinois

No investigation has been conducted by the FBI concerning the captioned individual. However, this Bureau's files reflect the receipt of a flier issued by the Council for Pan-American Democracy advertising the "Night of the Americas" to be held at the Martin Beck Theater on February 14, 1943, in New York City. The flier carried a partial list of sponsors and guests of honor which included the name of "Walt Disney."(u)

The Council for Pan-American Democracy has been designated by the Attorney General of the United States pursuant to Executive Order 10450.(u)

The "Peoples Voice," issue of January 15, 1944, contained an article captioned "New Masses Sponsors Tribute to Art Young." The article set forth that "New Masses" was sponsoring a mass meeting to pay tribute to Art Young, Dean of American Cartoonists who died recently. It was indicated that the meeting would be held on January 27, 1944, at Manhattan Center, 34th and 8th Avenue, in New York City. Among the individual sponsors of the meeting was listed the name "Walt Disney."(u)

According to the Special Committee on Un-American Activities in its report dated March 29, 1944, "New Masses" is a "nationally circulated weekly journal of the Communist Party."(u)(62-60527-25375)

The foregoing information is furnished to you as a result of your request for an FBI file check and is not to be construed as a clearance or a nonclearance of the individual involved. This information is furnished for your use and is not to be disseminated outside of your agency.(u)

RECORDED-112
INDEXED-112
62-60527-45803

Orig to USIA
Req. Rec'd. 4-25-55
W. L. Marshall:eah
(4)

12 MAY 10 1955

CONFIDENTIAL

DECLASSIFIED BY SP1GSK/KLS ON 5/16/80

Tolson
Boardman
Nichols
Belmont
Harbo
Mohr
Parsons
Rosen
Tamm
Sizoo
Winterrowd
Tele. Room
Holloman
Gandy

Figure 23. FBI summary of Disney's attendance at the "Night of the Americas" and the tribute to Art Young, provided in response to a request by the United States Information Agency, 1955. (Walt Disney's FBI file).

tell Franklin Roosevelt "that there was nothing illegal about being a Communist and that, therefore, the FBI had no grounds for information gathering."[17] In 1939, however, Roosevelt gave Hoover a directive so ambiguous that it allowed the Bureau to engage in a broad range of investigations not only of groups that *had* violated federal laws, including those on subversive political organizing, but that one day *might* violate them. Roosevelt made the FBI the clearinghouse for all information, from all law enforcement agencies, about these apparently subversive activities.[18] Then, in 1940, Congress passed the Smith Act, which made it a crime "to knowingly or willfully advocate, abet, advise, or teach the duty, necessity, desirability, or propriety for overthrowing any government of the United States by force or violence," and also criminalized any effort "to organize or help to organize any society, group, or assembly of persons who teach, advocate, or encourage the overthrow or destruction of any government in the United States by force or violence." The 1941 Voorhis Act insisted that all organizations deemed subversive register with the government.[19]

Disney became at least a partial victim of the government's bipartisan paranoia during the late 1930s and early 1940s. In fact, the FBI's slight cautions against Disney show the uncertainty within the government itself and the inability of the various branches of the government, during the period, to coordinate their activities. Almost certainly, Disney appeared at the Council for Pan American Democracy, and also the meeting of the League of American Writers (which had as a primary concern U.S. relations with South America), because the State Department had constructed him as something of an expert on Latin American affairs. By helping the State Department produce pro–United States propaganda for South American audiences, Disney became marked by the FBI as an unlikely, but nevertheless possible, fellow traveler of the Communist Party.

Throughout the 1940s and 1950s, Congress passed legislation (the Internal Security Act of 1950, for instance, and the Communist Control Act of 1954) aimed at curtailing the activities of groups deemed subversive. Communist-hunting became the raison d'être of the FBI, the best public relations tool that Hoover had to perpetuate popular support of the Bureau. As late as 1960, Hoover testified before Congress that "we now have 155 known, or suspected, Communist-front and Communist-infiltrated organizations under investigation," and that number kept growing. In 1962, Hoover told Congress that the number had gone up to "some

200."[20] The insistence of FBI officials that Disney's patriotism might still be slightly questionable, twenty years after they had gathered their last vaguely incriminating evidence, therefore worked for Hoover as one more indication that anti-Communist vigilance must never end, and that the threat of foreign domination of the United States still lurked everywhere.

At least as early as 1945, special agents for the Bureau had begun investigating what they referred to in their reports as "Soviet Propaganda in the Motion Picture Industry," which they classified under "Internal Security." Although much of this information has been censored, it appears that the FBI had some concerns that major figures in the U.S. film industry were being used to promote beliefs that the Soviet society was an open one and that the Soviet government was deeply concerned with the welfare of its citizens. A report from December 10, 1945, and covering the period from June through November, noted that "Russian delegates to the San Francisco Conference had asked Walt Disney to visit the Soviet Union to teach health and sanitation ideas through short film subjects." This must be a reference to a United Nations conference, and indicates that immediately after the war the FBI charted what it considered to be Soviet public relations efforts, and viewed Disney, through this proposed visit, as a potential unwitting pro-Soviet propagandist.

This concern emphasizes Disney's status as a cultural icon. His mere presence generated tremendous power; the State Department believed that his tour of South America in 1941 would help to establish United States commercial supremacy there, and this proposed trip to the U.S.S.R. made the FBI take notice. Despite its fear that Disney could be duped by the Soviets, and despite its suspicions about some of his appearances, however, the FBI also viewed Disney as a very willing anti-Communist, and enlisted him in their cooperative efforts with Congress to de-Sovietize the film industry.

Disney as Friendly Witness and FBI Propagandist

In an August 12, 1947, report labeled "House Committee on Un-American Affairs," the FBI listed Disney under "Possible Friendly Witnesses." The House Committee began in earnest its investigation of Communist influence in Hollywood in September 1947, when Committee Chair J. Parnell Thomas subpoenaed forty-five industry members of various political persuasions.[21] This FBI memo, however, shows that the Bureau was marshaling its forces to help Thomas before he went into action. As is common practice in FBI files, all names besides that of the subject of the researcher's interest are blacked out. However, judging by the size of the three columns of names and by the size of the typeface used in printing Disney's name, the FBI list of "Possible Friendly Witnesses" contains at least forty-five names.

Disney did indeed testify, as one of about two dozen initial friendly witnesses. That the committee chose him from a much larger list compiled by the FBI might indicate his perceived importance as a figure who could help mold public opinion. In fact, one of the tensions that emerges from the FBI files comes from the Bureau officials' suspicions about Disney's attendance at those dubious functions set against his active anti-Communism.

Three full years before Disney's HUAC testimony, for instance, the FBI took note of his participation in the formation of the Motion Picture Alliance for the Preservation of American Ideals (MPA), a collection of filmmakers determined to fight Communism. The FBI collected the Alliance's statement of principles ("Our purposes are to uphold the American way of life, on the screen and among screen workers; to educate, not to smear"), press clippings about the MPA, and the Alliance's own publicity. An MPA flier in the file lists Disney as the First Vice-President, and an article taken from a November 14, 1944, *Time* magazine authoritatively labels actor Rosalind Russell and screenwriter Dudley Nichols as "leftists," and Disney and director Sam Wood as "rightists" (fig. 24). Then, in a memo from September 21, 1944, headed "Communist Infiltration of the Council of Hollywood Guilds and Unions," the FBI constructed the melodrama of the fight against Communism: on one side, the Council of Hollywood Guilds and Unions, formed in response to the other side, the MPA, which itself had been created "to combat degrading influences within the motion picture industry."

This discourse makes it clear just which side the FBI supported. Disney is aligned with the good guys, as his name appears prominently on the attached MPA statement of principles, along with President Sam Wood and Second Vice-President (and MGM art director) Cedric Gibbons. Thus, even as the Bureau in the 1950s informed various government agencies of Disney's attendance at events sponsored by "subversive" groups years before, J. Edgar Hoover made Disney a SAC [special agent in charge] contact. In an office memorandum to Hoover from December 16, 1954, the Los Angeles special agent in charge wrote

OUR purposes are to uphold the American way of life, on the screen and among screen workers; to educate, not to smear.

We seek to make a rallying place for the vast, silent majority of our fellow workers; to give voice to their unwavering loyalty to democratic forms and so to drown out the highly vocal, lunatic fringe of dissidents; to present to our fellow countrymen the vision of a great American industry united in upholding the American faith.

These are our purposes. We have no others.

MOTION PICTURE ALLIANCE FOR THE PRESERVATION OF AMERICAN IDEALS

OFFICERS

SAM WOOD, President

WALT DISNEY, First Vice-President
CEDRIC GIBBONS, Second Vice-President
NORMAN TAUROG, Third Vice-President

LOUIS D. LIGHTON, Secretary
CLARENCE BROWN, Treasurer
GEORGE BRUCE, Executive Secretary

EXECUTIVE COMMITTEE

JAMES K. McGUINNESS, Chairman

BORDON CHASE
CARL COOPER
VICTOR FLEMING
ARNOLD GILLESPIE
FRANK GRUBER
RUPERT HUGHES
BERT KALMAR

FRED NIBLO, JR.
OSCAR T. OLDKNOW
CLIFF REID
WALTER A. REDMOND
CASEY ROBINSON
HOWARD EMMETT ROGERS
LELA E. ROGERS

HARRY RUSKIN
MORRIE RYSKIND
JOSEPH P. TUOHY
KING VIDOR
ROBERT M. W. VOGEL
GEORGE WAGGNER

Figure 24. Flier for the anti-Communist Motion Picture Alliance for the Preservation of American Ideals, listing Disney as First Vice-President (Walt Disney's FBI file).

that "because of Mr. Disney's position as the foremost producer of cartoon films in the motion picture industry and his prominence and wide acquaintanceship in film production matters, it is believed that he can be of valuable assistance to this office and therefore it is my recommendation that he be approved as an SAC contact." This memo makes it difficult to determine Disney's precise responsibilities as a contact. Under "Services Contact Can Perform," the Bureau office that monitors Freedom of Information Act requests has censored much of the information, yet has left in that "Mr. Disney has volunteered representatives of this office complete access to the facilities of Disneyland for use in connection with official matters and for recreational purposes."

At least strongly implied here are the possibilities for surveillance at Disneyland. Hoover seems to have been convinced of Disney's potential services, because later memos regularly referred to Disney as a contact, and in 1964 the film producer received one of several "letters of appreciation" sent out to various contacts. Disney himself apparently viewed his role as that of FBI propagandist, through his movies and his amusement park. According to a memo about his status as a contact, Disney "raised the question as to whether it would be possible to prepare a display or demonstration [at Disneyland] of how science is employed by the FBI in law enforcement." Thus, Disney offered the FBI a public relations opportunity to reach the millions who visited Disneyland each year. Alongside Disney's monorails, teacup rides, and roller coasters the tourists would see examples of the FBI's scientific law enforcement. The appeal must have been a powerful one, and the agent who wrote the memo asserted that Disney was "a very reliable individual . . . [and that] the Disneyland Amusement Park appears to have been popularly received." The memo also noted Disney's own concern that the FBI might be "reluctant to participate in any displays of a commercial nature where admissions are charged," and, indeed, FBI higher-ups summarily dismissed the proposal. Three of them added handwritten notes to the memo: "I don't see how we can do anything," followed by "I agree" and "I concur."

Although officials at the Bureau rejected the propaganda possibilities of Disneyland, they were by no means averse to Disney's functioning as FBI propagandist in other ways. Throughout the 1950s, Disney, who probably did not know about the agency's continuing concerns over his activities, offered to make films for and about the FBI, and the Bureau's negotiations with Disney construct a case study in institutional image making. An office memorandum from January 20, 1956, outlined Disney's plans as they were told to Agent Kemper. For his *Mickey Mouse Club* television show, the producer wanted to "take a group of children, if possible the children of Special Agents, and have two short scenes which would run about two minutes on the Mickey Mouse newsreel." The first would take place at a target range, and the other would explain "how fingerprints are taken." The agent writing the memo recommended cooperation with Disney, but the officials reading the memo disagreed, pencilling in: "I don't think we should," "I agree," and "OK."

One year later, Disney had not given up. In a March 1, 1957, letter from Hugo Johnson, Disney's newsreel representative, to Assistant Director of the FBI Louis B. Nichols, Johnson brought up the possibility "of stories regarding your organization." In response, a Los Angeles special agent wrote to Hoover that, "for some time . . . Mr. Disney has been interested in producing something featuring the FBI either for his *Disneyland* television show or the *Mickey Mouse Club* television production," and "that a show about the FBI with a Laboratory feature could be . . . produced in time for the fall, 1957 *Mickey Mouse Club* program."

The agent stressed that Disney's program had "an educational appeal," and then emphasized that the program had "an estimated audience of 18 million, Monday through Friday."

Hoover, understandably, must have been moved by the sheer numbers. An office memorandum from March 4, 1957, explained that the FBI Laboratory's 25th anniversary would be coming up in November, and that "some long-range planning is necessary if we want to take advantage of some of the better publicity media." The FBI clearly wanted to stress the laboratory because of its connotations of enlightened science and technology at the service of a benevolent government institution, but the memo still expressed some caution: "We, of course, will have to have more details as to commercial sponsorship." If the *Mickey Mouse Club* ran commercials from suspect corporate sponsors, the FBI could not appear on the show.

Within a few days, this concern over sponsorship seems to have passed, and FBI officials, just like television network executives, began thinking about audience and time-slot. On March 8, 1957, a memo insisted that "it is not felt that the *Mickey Mouse Club* is the proper place to publicize [the laboratory's] anniversary. It is a good show. It comes on at 5:30 P.M. each weekday and is aimed at the 'small fry.' If we are going to do this, we should do it right and try and get Disney to do a one-hour *Disneyland* show which at present is at 7:30 P.M. on Wednesday night. This show has an adult and juvenile appeal." The FBI wanted the prestige of prime-time for its propaganda, and envisioned indoctrinating not only young people but grown-ups into the scientific marvels of the Bureau.

The same memo shows a decided preference for cartoons over any other kind of film, saying that "by animation we could show ancient, medieval, dark ages and 19th-century law enforcement practices, the branding of the criminal, the dunking of the witches, etc. . . . Then using Sherlock Holmes with his magnifying glass and Sir Henry with his fingerprints we could begin to bring law enforcement up to date. Scientific law enforcement would reach its culmination in the Director's establishment of the Laboratory in 1932." Bureau officials wanted the FBI presented as the logical endpoint of an evolutionary process in which law enforcement became not only better and better but also more and more humane, with the laboratory representing everything that was both modern and benevolent in 1950s-style FBI practice.

The Bureau finally had to settle for live action over animation, for modern stories instead of the preferred historical one, and for presentation on the *Mickey Mouse Club*. For that program, Disney made four eleven-minute films about the FBI Laboratory, the Identification Division, and special agent training. Once again indicating the FBI's interest in as mature an audience as possible, a memo from October 18, 1957, asserted that the FBI was able to get its way on the issue of viewership, as "the Disney people are elevating the age plane of this series so that it will be of primary interest to youngsters in the 12- to 16-year age bracket."

An earlier memo stands out as the most revelatory of FBI thinking. The Bureau sought out information not only about possible Communist sponsors of Disney's television shows but also about the Communist connections of the FBI films' cast members. A thirteen-year-old, Dirk Metzger, starred in the films, and demonstrating that the FBI believed that surveillance could never start too young, and that, indeed, anyone could have contacts with the enemy, a memo from May 15, 1957, assured Bureau officials that FBI files "are negative regarding" the young actor.

Disney gave the FBI script supervision, and the Bureau made a number of changes that seem fairly innocuous, and which appear to be based on aesthetic judgment rather than on maintaining security or controlling the representation of the FBI. "This scene is out of place," the Bureau complained at one point, and, about another, said, "it is felt that the transition between the film on the Identification Division and the introduction of the Nazi spy case is a little rough." Then, after Disney completed the films, he wrote to Hoover to thank him "sincerely for the unstinted cooperation you gave to . . . our series." From these documents, relations between Disney and the Bureau appear to have been smooth.

Just before the films were to be shown, however, in January 1958, the strain between the Bureau and the producer became apparent. In several documents, FBI officials complained vigorously over not being shown final cuts of the films. One memo explained that the FBI "had been assured" of seeing a final cut, another stated that "we have naturally protested" Disney's reluctance to show the final films to the Bureau, and yet another succinctly asserted that the FBI "most certainly will take this treatment into consideration the next time the Disney Studios ask for cooperation."

Here we can see the reemergence of tension between the Bureau and Disney, between the government institution and one of its principal supporters and propagandists. Throughout the 1950s this tension grew, and it turned into an issue not only of formal control but also of representation. More precisely, the arguments between Disney and the FBI centered around image making and around who controlled the production of images of the Bureau.

The FBI learned from Hedda Hopper's column in the *New York Daily*

News from February 20, 1961, that Disney planned to make *Moon Pilot*, a live-action film. Because the movie concerned a Bureau agent, a memo advised "that discreet inquiries be made to determine the nature of the script and how the FBI Agent is portrayed." Later memos stressed that "efforts will be made to determine the contents of this film, particularly that portion portraying an FBI Agent," and that Los Angeles agents "made discreet inquiry at the Motion Picture Production Code Office to determine if script for this film has been received." By the end of February 1961, the Bureau had received a rather vague phone call from Disney, and a memo "assumed that Disney's contact with this office is to comment concerning" the film. Just a few weeks later, though, Disney and the FBI apparently still had not talked, and a polite but direct memo from Hoover to a Los Angeles special agent insisted that "you should arrange to personally confer with Walt Disney concerning his proposed filming of the story *Moon Pilot*. Tactfully point out to him the uncomplimentary manner in which FBI Agents are depicted. Advise him that the Bureau will strongly object to any portrayal of the FBI in this film. . . . Handle diplomatically" (fig. 25).

Bureau officials objected to the film's proposed depiction of surveillance methods, agent intelligence, and agent moral fiber. A memo from March 13, 1961, complained that "most references to the FBI are handled inaccurately and some are ludicrous. The Air Force officer . . . is continually outwitting surveilling Agents . . . at one point . . . the Agent in charge of the detail immediately arrests all [hotel] kitchen and dining room help. . . . One of the characters says that only one Agent saw her and the Agent thought the girl was a 'floozy' trying to pick the officer up. 'When she didn't he came back and made a play for her himself.'" Finally, the FBI believed that the film threatened the notion of a fluid chain of command, with resultant damage to Hoover's image as all-knowing overseer: "The principal FBI agent . . . pleads with the Air Force officer . . . 'If you'll tell me where the girl really came from I'll promise not to tell anybody, not even J. Edgar Hoover.'"

At the end of March, a Bureau agent visited Disney, who assured him "that if [the] Bureau objects he would change the script to eliminate the FBI and substitute another security agency. . . . He requested that Director Hoover review the script." That script, however, never came to the FBI, and so the Bureau had to be satisfied with a screening of the finished product, arranged "through the courtesy of the Air Force." In April, Agent C. D. DeLoach wrote the Bureau, saying, "It was my understanding that Mr. Disney had originally intended to portray FBI Agents in this movie, and he has done so to all intents and purposes, despite our

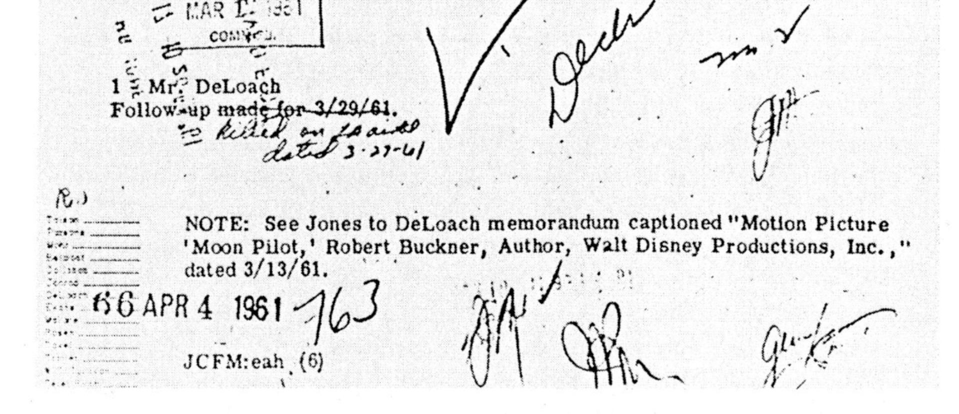

To: SAC, Los Angeles (94-761)

From: Director, FBI

"MOON PILOT"
MOTION PICTURE BY WALT DISNEY
RESEARCH (CRIME RECORDS)

Reurairtel dated 3/1/61.

You should arrange to personally confer with Walt Disney concerning his proposed filming of the story "Moon Pilot." Tactfully point out to him the uncomplimentary manner in which FBI Agents are depicted. Advise him that the Bureau will strongly object to any portrayal of the FBI in this film. As you will note from the story, FBI action basically involves guarding of the Air Force officer who is to make the first flight to the moon. Suggest to Mr. Disney that since FBI jurisdiction does not extend to the guarding of individuals that this action can be better represented by another Government agency. Handle diplomatically.

MAR 1961

1 - Mr. DeLoach
Follow-up ~~made for 3/29/61.~~
killed on LA airtel dated 3-27-61

NOTE: See Jones to DeLoach memorandum captioned "Motion Picture 'Moon Pilot,' Robert Buckner, Author, Walt Disney Productions, Inc.," dated 3/13/61.

66 APR 4 1961

JCFM:eah (6)

Figure 25. FBI memo of 1961 concerning representation of Bureau agents in a Walt Disney film, *Moon Pilot* (Walt Disney's FBI file).

protests, even though the Agents are not named as such. Needless to say, the Boss [Hoover] was amazed that Disney would produce such a picture which carries implications of criticisms of the FBI." The suspicion that Disney perhaps was poking fun at the Bureau constituted a genuine breach of faith and was perceived as damaging to the Bureau's relationship with the general public.

In describing Hoover's reaction, DeLoach may have been referring to a handwritten note on one of the reviews of *Moon Pilot* in the Bureau's clipping file. The film critic for the *Washington Daily News* wrote that, in the film, Disney was "leveling his humorous rifle at . . . members of Congress, the Air Force, the FBI, and French movie starlets," and then added that *Moon Pilot* proved "the FBI is . . . ineffectual." Even though Disney never referred to the Bureau by name in the film, the reviewer was not fooled about who the agents were. An FBI official reading the review (the signature, just an initial, resembles an "H" although it remains indecipherable, so that one cannot be certain that the official was Hoover), underlined Disney's name in the review, each mention of the FBI, and then the critic's assertion that "you may relish the sight of so many upstanding agencies and arms of the Government squirming thru a series of situations." The official then made a comment in the margin, one that was almost certainly tongue in cheek but that also indicated just how seriously the FBI defended its image as well as the Bureau's assumptions about anyone who might question that representation. "I am amazed Disney would do this," the official wrote. "He probably has been infiltrated."

When Disney considered putting *That Darn Cat* into production, the FBI protested that the co-writer, Gordon Gordon, "is a former Special Agent and has . . . capitalized considerably in his writings on his former affiliation with the FBI." A number of memos related Bureau frustration over Disney's refusal to provide officials with a script of the film, in spite of his promise to represent the FBI "as a most dignified and efficient organization." Showing the extent of the FBI's concern with image-making about the Bureau, agents made use of at least one studio spy to gain information about the film and to spirit a screenplay out of the studio. A memo referred to an "established source at the Disney Studios" who kept the Bureau informed about casting and production schedules but who "does not have access to a copy of the script for this picture." Thus, by the early 1960s, Disney the special agent contact, whom the FBI had trusted to expose Communism in the film industry, himself was being spied upon.

The Disney FBI file proves that, at least by the 1950s, one of the primary functions of the Bureau was protecting its image. The Bureau devoted extensive effort to the task and even engaged in studio espionage. Moreover, Disney's file from the 1940s until his death serves to complicate our notion of the relationship between the Bureau and the individual or, at least, the more celebrated individual. We tend to think of the FBI in an adversarial relationship with writers such as William Faulkner and Howard Fast, and in a sympathetic relationship with such right-wing figures as John Wayne and Disney. But the Bureau's information on Disney shows that these relationships were never so simple, and that, in Disney's case, the FBI constructed several contradictory incarnations of the filmmaker. He was a chief propagandist but also a producer of potentially subversive movies. He served as a friendly witness and as a contact for agents, but, as I have indicated, the Bureau also thought of him as a possible Soviet dupe. Thus, the FBI came to specialize in image construction, creating private citizens in various images while jealously guarding the right to create an image of itself.

The Treasury Department: Disney Animation, Tax Collection, and Audience Reaction

FBI documents about Disney show the contradictions within a single government organization over a quarter-century. The Treasury Department file on Disney, however, reveals the tensions between separate institutions—Treasury and Congress—and, of even more interest, helps us reconstruct the film audience at a specific moment in history.

In 1942, Disney complied with a Treasury Department request to make a cartoon about the country's new tax laws, laws that required millions of citizens from the lower and lower middle classes to pay taxes for the first time. The eight-minute cartoon, called *The New Spirit,* starred Donald Duck. At the beginning of the film Donald balks at paying his taxes, but then, because of a radio voice that conjures up images of the threat posed by Germany and Japan, Donald realizes that, for the average citizen, complying with tax laws during a time of war serves as the highest form of patriotism.

The Donald Duck Debate

The New Spirit might seem like a neutral enough collaboration between the government and Disney, especially considering the much larger-scale joint projects between Disney and the State Department.[22] The Treasury Department file about the film, however, reveals the disagreement within the government over just how to propagandize the new tax laws and precisely how to pay for that propaganda. A portion of the

Congressional Record from February 9, 1942, has been placed in the file, and it tells the story of California Senator Sheridan Downey's defense of Donald Duck.[23] Downey began his speech in the Senate by criticizing those "detractors" who were blocking the appointment of actor Melvyn Douglas to the Office of Civilian Defense because of their belief that Douglas was a Communist.

Representing California, Downey clearly had a stake in maintaining the good name of the motion picture industry, and so after his homage to Douglas, he moved on to Disney's film. About the cartoon star, Downey claimed that "poor Donald . . . is now in the doghouse, placed there by a slim majority in the House of Representatives." This remark referred to the House's refusal to pay the $80,000 it cost to produce *The New Spirit*, with Downey adding that "for only a portion of the out-of-pocket money a film of incalculable propaganda value was given our government. . . . I hope Mr. Disney and his associates will not believe that the Government of the United States and the people of this Nation are as ungrateful and as unappreciative as the vote in the House of Representatives would seem to indicate." The problem seems to have been solely over money. Downey's speech implies that the term propaganda had not acquired all of the negative connotations it has today, and that the need for propaganda, in this case to persuade low- and middle-income citizens to pay taxes, could not be questioned.

The recalcitrance on the part of the House put the Treasury Department in something of a bind. In anticipation of congressional approval, a December 19, 1941, memo to Treasury Secretary Henry Morgenthau shows that a deal with Disney had already been made: "Walt Disney agrees to produce and deliver . . . 700 to 800 feet of film in technicolor with a tentative title *The New Spirit* at a cost not to exceed $40,000," with another $40,000 to be spent on additional prints of the movie.

Performing some legal gymnastics to justify the government's payment to Disney, E. H. Foley, the general counsel to the Treasury Department, explained the important role propaganda played in the government's plans and something of the perceived responsibility of the government to its citizens. In a letter to Morgenthau on February 11, 1942, Foley stated that the point of *The New Spirit* was to "stimulate public interest in the payment of income taxes and to enlighten the public with respect thereto." Clearly, while such members of the government as Downey and Foley could approvingly refer to the film as propaganda, they also had to make it clear that the propaganda was absolutely benevolent. It was, after all, only educational, its intent being to "enlighten." This is language similar to that used by the State Department to justify Disney's film projects for and about South America and indicates a similar disingenuousness on the part of government officials. Disney's films for the State Department were designed not only to educate but to make U.S. presence in South America appear beneficial. In the same manner, *The New Spirit* was not designed so much to "stimulate public interest" as it was to appeal to a certain aspect of that public, the lower and lower middle classes, and to assure them of the patriotism of paying taxes.

Foley later acknowledged the class divisions in the audience, and indicated that, because of them, it was the government's responsibility to make the film. "When Congress," he wrote, "enacts legislation for the taxation of a large class of persons who have never before been subject to taxation, it seems manifest that it is the duty of the Commissioner of Internal Revenue and the Secretary of the Treasury to take steps to advise such taxpayers of their liabilities and duties." He added that "should those officers fail to do so, I am constrained to believe that they would be derelict in the performance of their duties." Foley may or may not have been stretching for a legal point here. What seems certain, however, is the assumption of the unique power of film to reach a huge audience and to shape that audience's behavior, and also the government's belief that, because of this power, the federal bureaucracy must have access to the medium.

The Heroism of Production and Institutional Cooperation

Congress never agreed with Foley and refused to pay for the film. That decision meant that the Treasury had to foot the whole bill, and so the department took on the task of convincing U.S. citizens of the worth of the cartoon. The Treasury mounted a large publicity campaign and carefully monitored all of the information about *The New Spirit* that reached the public. A Treasury press release from January 23, 1942, announced the agreement between Disney and the Treasury, and claimed that Disney and his workers were "setting new production records to have [the film] ready for showing within the next ten days in 12,000 theaters throughout the country." The department's strategy was to make the very production of the film heroic, as if that would indicate the necessity for the production itself. The Treasury also depended on star power, selling the picture by selling its best-known names. The last lines of the release explained that "Treasury tax experts called upon Disney for guidance in determining Donald Duck's status and decided Donald was the 'head of a family' because of his support of his three adopted nephews, 'for whose maintenance he has a legal and moral obligation.'" Clearly, Disney became the central star of the film's production, guiding those tax experts, and Donald became an ethical force. Even more constructed than most Hollywood stars—Donald existed only as ink and paint—he nevertheless emerged from department publicity as a role model, as a patriarch whose highest duty was to his dependents.

On January 27, another release exulted in the film's completion, and the point of the publicity once again was the miracle of the movie's production. "Walt Disney broke all of his production records," the release explained, "leading his crew of artists and technicians in working around the clock seven days a week, in order to get the picture set for a maximum number of showings before the income tax deadline." Disney became the hero in this story of his film's production, and, at least implicitly, the Treasury posited any sacrifices required of the taxpayers as insignificant compared to those suffered by Disney. The corporate magnate, like Donald, emerged as role model. For the Treasury, as important as the film's propaganda possibilities was the propaganda potential of the facts of the film's production.

Along with these releases, the Treasury, on February 20, 1942, sent out photographs from the film to newspapers across the country and worked closely with print journalism and Hollywood studio production departments to generate publicity and to monitor it. Charles Schwarz, the Treasury's director of public relations, wrote a note on January 29, 1942, thanking the *New York Herald Tribune* for publishing photos "from the Donald Duck picture." George Buffington, the assistant secretary of the treasury, received a note written on January 30 from Martin Quigley, publisher of the *Motion Picture Herald*, along with a copy of the week's issue of the paper, "which contains . . . a reproduction of Mr. Morgenthau's letter, together with article and illustrations, on the Disney film." On January 27, Schwarz wrote to Disney publicist Antoinette Spitzer that "the *Life* [magazine] people called me this afternoon to say that they had scheduled two color pages for the issue out on March 13. I pointed out that some black and white before then would also help and they promised to see whether New York could make an exception and handle the story twice."

These notes back and forth imply that the government enjoyed a cozy relationship with print journalism, at least around issues concerning the war and patriotism (the film advises paying taxes "to beat the Axis"). *Life* could be convinced to run more stories if the Treasury requested them, and newspapers eagerly kept the department informed of their efforts to publicize the government's movie. A similar relationship, but not necessarily for similar reasons, appears to have existed between the government and the Hollywood studios. On February 19, 1942, for instance, Frank La Falce, Warner Bros.' director of advertising and publicity, sent Schwarz "tear sheets of breaks on the theatre pages of Washington papers on the Disney short subject, *The New Spirit*." On February 2, Schwarz wrote to Monroe Greenthall of United Artists, assuring him that "if there is anything further we can do on this end to collaborate with your plans [to publicize the film], I shall be glad to hear from you."

Neither Warner Bros. nor United Artists distributed Disney films at this time (RKO did). Warner Bros., of course, could have made some money from the film by showing *The New Spirit* in the company's theaters. United Artists, which owned very few theaters, could have received little economic benefit from the film. But these studios' efforts on the government's behalf might well indicate Hollywood's continuing concern to conciliate. The Justice Department already had taken antitrust action against the movie studios, and, while the threat of government censorship had been lessened by the self-imposed stiffening of the Production Code in 1934, the studios probably felt it still necessary to take pains to convince the government that the film industry did not have to be policed. Just before Pearl Harbor, Congress investigated whether there was a discernible bias toward any of the warring nations in Hollywood films, and just after the attack the government sought to enlist the movie studios' support in the war effort.[24] Indeed, during this period, the relationship between the government and the film industry was a tense one, with the latter determined to maintain its artistic and economic integrity without antagonizing the former. Despite this uneasy balance, the publicity documents in the Treasury file demonstrate how apparently independent and frequently antagonistic institutions—journalism, the government, and motion pictures—actually worked together to consolidate public opinion about the new tax laws.

The Taxpayer as Movie Critic

To make sure of that opinion, the Treasury Department instructed Internal Revenue Service collectors and deputy collectors to report on the film's reception. Most of the reports are drily informational. On February 19, 1942, for instance, John P. Malik wrote the Treasury that "as far as the audience was concerned, it appeared that the reaction was favorable." Another reaction report, from the same date, said that the theater manager "stated that the film was very popular." A few of the reports, however, provide us with more information about the film audience, explaining its numbers, expectations, and divisions.

On February 20, Deputy Collector John J. Drabicki reported that in one week at a major theater, "37,261 movie patrons shouted enthusiastic approval of the movie short." He added that the film's information about taxes "creates . . . patriotism," and that the "public wishes to see more movie shorts of this type." These numbers are certainly of interest in telling us how many people might attend a movie theater in a given week (from the report, the location of the theater is unclear), while Drabicki's assertion about the effectiveness of the propaganda, if accurate, might make us wonder about the demographics of that audience. Who, precisely, "wishes" to see more of these shorts?

A partial answer comes from a Chicago report. On February 19, a deputy collector wrote that he had been "requested to attend the Elmo Theatre, located at 2406 West Van Buren Street . . . to observe the reaction of the public to the presentation of Walt Disney's Movie Short, 'Donald Duck Filing His Income Tax.'" He prefaced his report on the audience's response by explaining the audience's class: "This is one of the poorer sections of the city, with average incomes ranging from $800.00 a year to $1800.00 a year." These people constituted the crucial audience for the film, because the new tax laws turned them into taxpayers, and the collector reported that "the picture was enthusiastically received." He then explained the precise audience identification with the star of the film. Viewers closely followed "the various reactions of 'Donald Duck,'" he wrote, adding that "when Donald groaned at the mention of Income Tax, the audience groaned; when he decided to do his bit by paying it, they cheered him."

The deputy collector realized that not only was the class of the audience a crucial issue when discussing the film, but that the star system was, too. At least since the widespread acceptance of that system by the film industry between 1910 and 1915, Hollywood had manufactured stars to sell any number of products.[25] These audience reports about *The New Spirit* demonstrate that the Treasury took advantage of that system, which functioned in animation just as it did in live-action films, to sell another kind of product, that is, government policy.

A report from February 20, 1942, by Deputy Collector Guy DeFilippis, indicates that the government film did indeed have a sales pitch. DeFilippis wrote that the "first announcement of Donald Duck feature brought usual reaction to Disney films, that of laughter and relaxation." He added that "this worked to set the audience in a receptive mood for income tax propaganda." Like the previous report, this one stressed the importance of Donald's star power. The closeup of him that started all of his films from this period—the "first announcement of Donald Duck feature" that DeFilippis mentioned—set in motion a whole series of pleasurable expectations. Clearly, DeFilippis approved of those expectations being not so much subverted but used to make the audience accept an entirely different kind of Disney cartoon, an overt propaganda piece.

The audience, however, noticed the "change of mood and tempo" and "followed [it] very closely." The first part of the film "gave the audience the feeling that the story was more propaganda than income tax instruction," and so betrayed that early "laughter and relaxation." These early scenes are marked by an entrance of exhortation and real-life concerns rare for a Disney film of the period; a voice from the radio harangues Donald about the war, and insists that taxes make it possible "to blast the aggressor from the seas" (fig. 26). Besides pointing out the failure of the opening segment of the cartoon to meet the audience's expectations, DeFilippis also suggested that the film created a split between Disney and the spectator. The former sought to create a "receptive mood for income tax propaganda," while the latter rejected propaganda as unsuitable for a cartoon. DeFilippis assured the Treasury, however, that "the last part . . . was instructional and all thoughts and feeling of propaganda were sidetrucked [*sic*]." According to this report, it was the star system that at first made the propaganda palatable. Then, because the propaganda deviated from the expectations generated by that star, the film began to lose its audience. Finally, because the propaganda became invisible, that is, seemed simply "instructional," the film succeeded. These documents about the reception of the film, therefore, frankly admitted that *The New Spirit* was a propaganda film. But they also made it clear that, for the film to succeed, the audience could not be able to acknowledge it as one.

At the end of his report DeFilippis described the audience's reactions to the entire cartoon. Although the film succeeded in "confusing some" while "making it clear to some that they could or could not use short [tax] form," *The New Spirit* achieved its propaganda purpose because it "aroused interest in income tax," and "definitely connected income tax revenue with the war." Finally, there was unanimity about the film, as DeFilippis could assert that "All like the 'short' and 'got a kick out of it.'"

Figure 26. "In Walt Disney's THE NEW SPIRIT, made under the auspices of the United States Treasury Department, Donald Duck rolls up his sleeves to work on convincing Mr. and Mrs. Taxpayer that they can and should pay Uncle Sam this year with a grin. In this scene from Donald's celluloid contribution to patriotism, he hears the radio's voice proclaiming the birth of a new spirit in America. Donald snaps to attention, gives a snappy salute." (Publicity still from *The New Spirit*, courtesy of the National Archives).

In addition to these department reports about the reception of the film, the Treasury file also contains unsolicited letters written to the department by some film viewers. These letters force us to question DeFilippis's belief in audience unanimity, and demonstrate that in analyzing reception in relation to historical context, film scholars must stress contradiction and disagreement rather than consensus. The letters show the heterogeneity of, and demonstrate the difficulty of ever reconstructing, the early 1940s film audience. Numerous viewers criticized *The New Spirit*, but they did so for a variety of reasons, and when they approved of Disney's film they hinted at the class divisions within the audience.

On January 26, 1942, Alfred T. Sihler, the assistant vice president of the Federal Reserve Bank of Chicago, wrote to George Buffington, the assistant to the secretary of the treasury, about his enthuasiasm for the cooperative effort between Disney and the Treasury. He then indicated that popular culture was just the place for the kind of instruction promised by *The New Spirit*, because "today we have the release on Irving Berlin's income tax song" (fig. 27). A few weeks later, on March 2, Jack D. Tarcher, of the New York-based J. D. Tarcher & Co. advertising firm, wrote to Secretary Morgenthau to "permit me to express my admiration for the good sense the Treasury showed in having prepared and released the Donald Duck movie on Taxes to Beat the Axis." Tarcher added that "as one whose tax bill is in five figures I can speak with some feeling on the subject. I have always been glad to pay my taxes and gladder still after Donald's exposition."

Another appreciative letter, from March 11, came from the Tax Institute at the University of Pennsylvania's Wharton School of Finance and Commerce. Others came from a physician in Monroe, Michigan; from the president of a New Jersey chemical company; and from fashionable New York addresses: 30 Sutton Place and 161 East 91st Street. These letters at least imply that the most widespread approval for the film originated in the middle and upper classes; people in fancy areas whose tax bills were substantial and who had comfortable jobs. Indeed, this might make sense for several reasons. As explained above, the new tax laws, while increasing taxes paid by the wealthy, were designed primarily to include large numbers of lower-class taxpayers for the first time. Longtime taxpayers, therefore, might have approved of a law that seemed to distribute the tax burden to all classes.

This should not be taken to mean that the well-to-to were simply fed up with the poor not paying any taxes. Most of the letters implied that, during wartime, it was everyone's patriotic duty to pay his or her share, and there are at least some approving letters that cannot be identified as coming from a member of the middle or upper class. A woman from Tampa, Florida, in a handwritten note on plain white paper (the other letters were typed on stationery), explained to Secretary Morgenthau that if *The New Spirit* "doesn't wake up every tax paying adult throughout the states then they cannot be one hundred percent Americans."

Taken together, though, these congratulatory letters imply a class basis to approval of the film. These letters might demonstrate that conceptions of patriotic duty, and of the government's role in telling the

January 26, 1942

Mr. George Buffington
Assistant to the Secretary
Treasury Department
Washington, D. C.

ans Jan. 28

Dear George:

I understand you are out in California conferring with Walt Disney on the tax cartoon, and today we have the release on Irving Berlin's income tax song. Looks like you will be rivaling Darryl Zanuck, and your friends will be coming to you to get passes for the studios!

We have had our nose to the Defense Bond grindstone so steadily that we have not been able to do much on the Tax Notes excepting take care of the orders and supply application forms, etc.

Hope you will be coming by Chicago after you have taken care of your Hollywood territory.

With best regards,

Sincerely yours,

Al

Alfred T. Sihler
Assistant Vice President

B

Just received your new tax note folder – & it looks fine –

Figure 27. "I understand you are out in California conferring with Walt Disney on the tax cartoon . . ." Letter in support of *The New Spirit* from the Federal Reserve Bank of Chicago (National Archives).

governed what that duty was, might have varied from class to class, although the letter from Tampa cautions us that we cannot simply view class opinion as monolithic, as something that was absolutely predictable within any single group of people.

Letters criticizing the film far outnumber these few supportive ones, and a clear-cut class analysis of the letter-writers is difficult. As opposed to the letters that approved of the movie, however, far more were handwritten than typed and very few were on stationery. More of the letters came from small towns than from big cities, and the grammar and writing in many of them indicate less education than might be inferred from the supportive letters. The critical letters most often dealt with the money spent to make the film, which might tell us something about class—the writers, already feeling disfranchised, resented government expenditure on extraneous items. But the criticism also focused on racism, the proper use of the flag, and the nature of patriotism.

Two of the letters point to a class antagonism toward the government's film. M. R. Brant, from Bethesda, Maryland, wrote on February 12, 1942, about conversations with "a lot of John Q. Publics," and provided reasons why so many of them hated the film. Brant wrote that the average citizen's "boy has to go to war and has a first class chance of being killed or mangled for a big sum of 21 dollars per which when the Officers get thru collecting his expenses from him he has a minus quantity left." The underlined "officers," the emphasis on the well-off taking money from the poor, and the stress on the latter's sacrifice—"being killed or mangled"—make it reasonable to interpret this as a letter from someone who, in some form or another, had personal experience to go by. Brant then complained about the "big fat salaries" of congressmen, and finished by asserting that, with the film, "we have a (Donald) duck at $80,000 a course."

In slightly more measured tones than Brant's, M. C. Bryson, from El Paso, Texas, explained that he/she was "one of the army of Americans who will pay an income tax for the first time this year." Bryson added that "I have noted your statement . . . that you consider the eighty-thousand dollars paid for the Donald Duck film to be a good investment." Having to pay taxes for the first time almost certainly placed Bryson in the lower middle or lower class, and, like Brant, the main point of the letter was to ask why "Mr. Disney [should] need . . . *our* eighty thousand dollars?

Referring to the cost of the film, other letters asked, "Has official Washington gone mad?" and "What the hell's wrong with you?" and "What a chump you were to let Disney stick you eighty thousand bucks for that Donald Duck short" (figs. 28 and 29). Indeed, in these angry letters the most common complaint concerned the money it cost to make the movie.

There were, however, other criticisms. For instance, two letters complained about the film's racism. J. G. Butler, the minister of the Summerfield Methodist Church in New Haven, Connecticut, wrote to Secretary Morgenthau on February 10, 1942, that, while *The New Spirit* "got its point across . . . unfortunately the film degenerated to a plane of unadulterated hate, with raucous voice and beastial [*sic*] figures representing Japanese and Germans." Butler then asked "that you would use your influence to see that such venom does not become part of the program of the Treasury Department in its efforts to get people to buy defense bonds."

June Hoffmann, writing on February 9 from the University of Connecticut, insisted that "I do not think that our government should participate in this type of hate-producing propaganda." She complained that, "in particular, the portrayal of German and Japanese men as beasts, the fiery scenes of destruction, and the snarling voice of the commentator will not arouse the kind of emotions which will seek a just and durable peace when this war is ended. In short, this picture is not worthy of our government."

From just two letters it is difficult to determine how widespread these sentiments were. Both writers had an institutional affiliation, a church and a university, so it is perhaps safe to assume that their ideas were representative of those of a larger group. In any case, Butler's letter and Hoffmann's demonstrate that, just a few months after the United States' entry into World War II, there was already criticism among those in the film audience concerning Hollywood's representation of the enemy.

The highest levels of government turned racism into official policy just a few days after the critical letters about *The New Spirit* arrived at the Treasury; on February 19, 1942, Franklin Roosevelt signed Executive Order 9066, which placed in internment camps all Japanese Americans living on the West Coast. Film historian Bernard F. Dick has claimed that "since the first anti-Nippon films did not reach the screen until the spring of 1942, they could not have contributed to the internment decision."[26] Obviously, *The New Spirit* was full of "anti-Nippon" sentiment, and that film appeared very early in 1942. Disney's film certainly did not contribute to the decision to remove Japanese Americans from the West Coast. But the film does indicate that the federal government attempted to generate anti-Japanese sentiment almost immediately

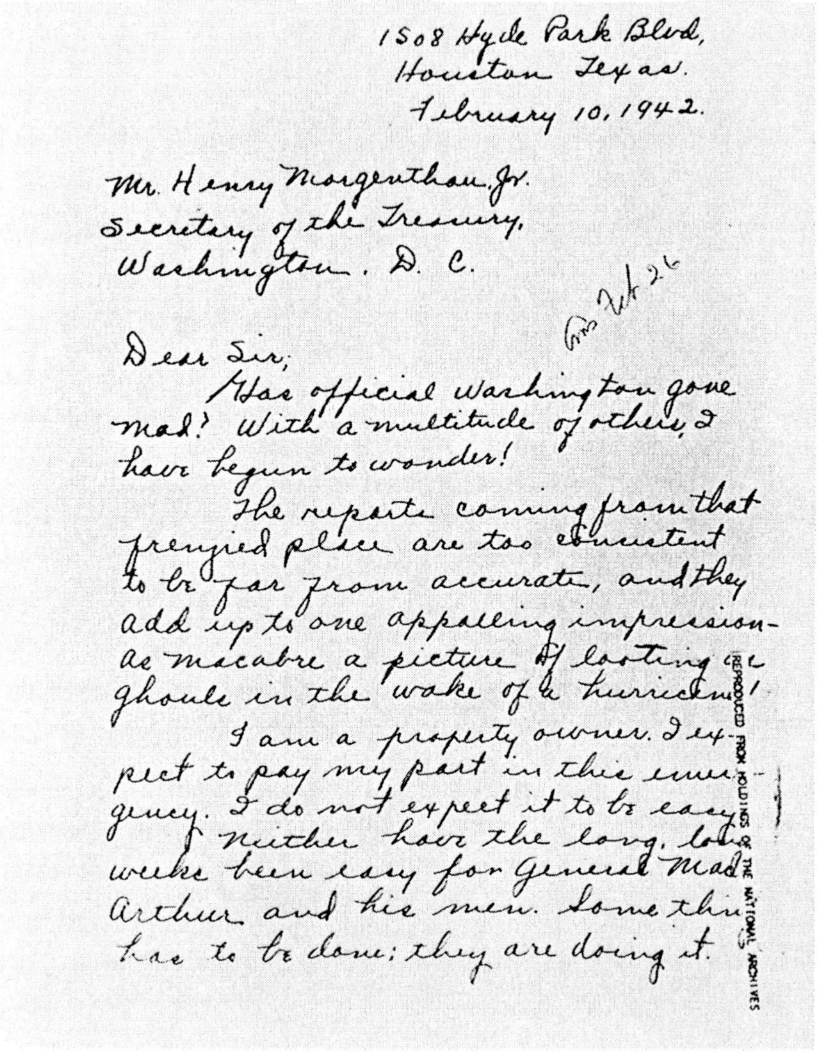

1508 Hyde Park Blvd,
Houston Texas.
February 10, 1942.

Mr. Henry Morgenthau, Jr.
Secretary of the Treasury,
Washington. D. C.

Dear Sir;
Has official Washington gone mad? With a multitude of others, I have begun to wonder!
The reports coming from that frenzied place are too consistent to be far from accurate, and they add up to one appalling impression—as macabre a picture of looting ghouls in the wake of a hurricane!
I am a property owner. I expect to pay my part in this emergency. I do not expect it to be easy. Neither have the long, lo[ng] weeks been easy for General Mac-Arthur and his men. Some thin[g] has to be done; they are doing it.

Figure 28. "Has official Washington gone mad?" Letter criticizing *The New Spirit* (National Archives).

Figure 29. "Sir—After reading this evenings [*sic*] paper, what the hell's wrong with you?" Letter criticizing *The New Spirit* (National Archives).

Feb 12-42
202 W. 2nd St.
North Manchester Ind.

Henry Morgenthau,
Washington, D.C.,

Sir:- After reading this evenings paper, what the hell's wrong with you? You'd throw-away $80,000 of our tax-payers hard earned money. If your salary compared to anything like mine, perhaps you could understand the value of it. My God man $81,000 is lots of money to throw away but you rich parasites who know nothing of frugal living but who are used to champayne banquets and luxurious vacations and the spending and wasting public money, haven't the brains of Donald Duck. Hope to Christ you come to your senses soon, its disgusting.

Yours Truly,
Karl L. Young.

THE WHITE HOUSE
WASHINGTON

February 14, 1942.

Respectfully referred to the Treasury Department for consideration and acknowledgment.

13 1942

WHO LIVES AT
IE OTHER NIGHT
AX SHORT IT
ERY NEXT DAY

after Pearl Harbor, while the critical letters imply that, at least early in the U.S. war effort, there was no public consensus about the proper depiction of the Japanese or the Germans.

Other letters questioned whether the film promoted patriotism properly. Almah White, who in her February 10 letter to Secretary Morgenthau asked, "Has official Washington gone mad?" explained that "I expect to pay my part in this emergency" and that she did not "expect it to be easy." White continued that "neither have the long, long weeks been easy for General MacArthur and his men. Something has to be done; they are doing it. The stimulus comes from within, not from Hollywood."

For White, being prodded into patriotism was no patriotism at all. She also resented being addressed as a child, asserting that "I think it is a national insult to imply that a country of adult taxpayers can be cajoled by a childish amusement. If it were true, I should question whether we had anything left for which to fight!" In 1942, animation had not yet been completely ghettoized as children's entertainment. That would come in the late 1940s and throughout the 1950s when animation became a television staple, aired early in the morning for children getting ready for school, or between three and five o'clock in the afternoon when children were coming home. However, White's remarks indicate that, by the 1940s, animation nevertheless strongly connoted kids' fare.

There were reasons for this. The movie studios had to create products for the broadest audience possible, and they did this through a system of differentiation on the standard movie theater bill. The typical afternoon or evening at the movies in the early 1940s consisted of a newsreel, a cartoon, at least one feature film, and also coming attractions and, perhaps, a live-action short subject. The newsreel was the adult entertainment, the cartoon was at least partially geared for children, and the feature film had elements that all would like.

Despite White's complaint, though, it made sense for the Treasury to make a cartoon instead of a feature-length, live-action film about the tax laws. A seven- or eight-minute animated short cost far less than a longer, live-action film and could be completed in much less time. If White was correct, however, and cartoons were considered more childish than adult, the making of *The New Spirit* indicated, at best, an attitude of paternalism and, at worst, one of contempt on the part of the government for the millions of lower-income citizens who would be paying income taxes for the first time. Indeed, forcing us to return to the issue of class in discussing the film, the unasked questions in White's letter were whether the government would even make a propaganda film directed primarily at the upper class, and, if so, would it be a cartoon?

In the same year that Disney produced *The New Spirit*, the federal government adopted the American Flag Code, which prohibited use of the flag in advertising or on any disposable item. In keeping with this heightened wartime interest in proper usage, two letter-writers referred to a representation of the flag in *The New Spirit* that they felt subverted the very patriotism that the film attempted to foster. One of the publicity photographs distributed by the Treasury showed a closeup from the film in which Donald Duck has American flags in his eyes (fig. 30). On February 11, 1942, Mary Turlay Robinson sent a reproduction of the photograph, along with a letter, to the "Arbiter on Uses of the Flag of the U.S.A." at the Library of Congress. She complained that "surely this constitutes a misuse of our flag, and should be dealt with accordingly. With the proposal to show this film nationally, it is far more serious than the patriotic but mistaken zeal of private persons who from time to time have placed the flag in positions on cars, etc. where it should not be." Archibald MacLeish, the librarian of Congress, sent Robinson's complaint on to the Treasury.

Similarly, on Febuary 10, Bartlett Burleigh James, who, in a handwritten addition to his letterhead, referred to himself as "Pres. The Cultural Institute of the Americas," complained of the same picture of Donald. He wrote that, "without discussing for a moment the authority resident in . . . the government to make a business-trade of the flag with an amusement concern . . . I am sure that I express the sense of many in deploring this as a misuse of the 'colors' under which are [*sic*] sons are fighting." At least in the early days of United States involvement in the war, then, there was no agreement over the use or meaning of those symbols—the evil Japanese and Germans as well as the United States flag—which we have come to assume represented the American cause for a majority of U.S. citizens.

To its credit, The Treasury responded to almost all of the complaints. Each of the responses is a little different, depending on the criticism, but the nature of each is similar. In at least two of the letters, both from February 14, 1942, a department official explained that "while we have noted your reason for the objection, I should like to point out that at this time it is most important to arouse a national realization of the desperate nature of the emergency which confronts us. Our adversaries are bitter and implacable." This explains the strategy of the Treasury: to build consensus, to create "a national realization."

Figure 30. "In this scene from Donald's celluloid contribution to patriotism, his 'zeal' shines in his eyes as he hears a voice on the radio proclaim that the 'new spirit' is that of a free people united in a common cause." (Publicity still from *The New Spirit*, courtesy of the National Archives).

Indeed, this was not very different from the goals of the State Department in its South America propaganda, or of the FBI in its concern with the image of the Bureau that reached the public. However, the film viewers' letters in the Treasury file show how impossible it was to create that consensus. The movie audience, which film studies has yet to deal with in sophisticated ways, was splintered, enjoying films for as many reasons as it hated them. At least during the 1940s, Walt Disney as an unproblematic national icon of artistic benevolence, as the popular journals created him, or as a symbol of patriotic consensus, as the Treasury hoped to construct him, simply did not exist for the general public.

AFTERWORD

Cartoons came to their audiences as those films most removed from reality, the ones most prone to magical transformations and remarkable characterizations. A variety of cartoon "experts" only served to encourage this understanding of the medium. Warner Bros. producer Leon Schlesinger, in the *Look* magazine article on censorship, stressed the importance of cartoons as fantasy entertainment for the child audience. Film critics considered animation to epitomize the marvels of Hollywood production. Those in charge of constructing the film bill used cartoons as the perfect complement to the more real-life concerns of the feature film and the newsreel.

I have tried to show that cartoons also served more overtly political ends. Carolyn Marvin, in her study of the cultural and political changes wrought by the widespread introduction of electricity at the end of the nineteenth century, has written that "communicative practices always express social patterning, [and that] any perceived shift in communication strikes the social nerve by strengthening or weakening familiar structures of association."[1] The animated short subject, as developed by Hollywood, marked just such a new communicative practice, and resulted in subtle shifts in the manner in which leisure pursuits were presented to, and enjoyed by, the public. Demonstrating a particularly modern concern with collapsing spatial and temporal differences in favor of wholeness, animation, and its place on the film bill, helped the cinema show, apparently, *everything* to its spectators. Then, demonstrating its own ability to cross borders, animation worked during World War II to bring United States ideology to various parts of the world, both to the frequently disgruntled American soldiers stationed overseas and also to indigenous populations. Moreover, the discourse about animation in a variety of popular journals during the 1940s functioned as part of the era's project of nationalizing belief systems through print medium practice, with *Time* magazine and the like supplanting more regional journals. Thus, both the federal government and a variety of corporations adapted animation to their own needs and projects, using a new entertainment medium to assert "traditional" values in the midst of rapid social change.

After having provided this monolithic meaning for cartoon practice, in which animation works relentlessly for the preservation of a conservative status quo, I must now warn against just such a formulation, even while acknowledging its staying power. Indeed, in the 1990s, a variation of it has become the fashionable way for the popular media to discuss Hollywood cartoons. Now, rather than serving the needs of conservative institutions, they are seen as expressing univocally, and in easily intelligible form, political messages that reflect the feelings of the masses. In February 1992, for instance, a New York theater presented "Cartoon Soup Kitchen," a collection of animated shorts from the 1930s that used familiar characters to portray the economic disaster of the Depression. No less an arbiter of such things than the *New York Times* referred to the films as typical of "the socially conscious cartoon" that dominated production during the era.[2] Reporting on a collection of animation art (posters, sketches, etc.) from the World War II period, the *Los Angeles Times* insisted that, like the movie stars who "toured with the USO . . . Hollywood's animated characters served the cause with equal fervor." The *Times* continued that "Donald Duck learned how to file his return under the newly enacted income tax laws . . . in *The New Spirit* . . . [and] gave new meaning to the Good Neighbor Policy in *Saludos Amigos*," while "Private Snafu entertained the troops in *The Army-Navy Screen Magazine*."[3]

In part responding to this belief in monolithic institutional power and unproblematic practices of interpreting cultural artifacts, John Fiske has written that popular culture is always political, because "it is produced and enjoyed under conditions of social subordination and is centrally implicated in the play of power in society."[4] Following Fiske, I have told a story of how competing groups, throughout the classical period, fought for control of an aspect of film practice—animation—as if they were aware of that practice's potential power. Within the film industry, producers struggled with censors over cartoon content, for instance, and with each other over audience segments and merchandise markets.

Within the government bureaucracy, the FBI fought with Walt Disney over the proper representation of authority. And always, there was the implicit conflict between film text and film audience. Copyright documents, for example, indicate the possibility for multiple readings of cartoons, despite their seeming monolithic constructions of gender or race or class. In addition, the very determination of the 1930s film bill to make differences disappear testifies to the existence of differences during that period. A few years later, the cartoons produced during the war either directly addressed dissension in the enlisted ranks, or were the occasion for its expression in domestic audiences angered by the attempted creation of a univocal patriotic consensus.

Thus, animation was not produced within a system of fixed institutions and social practices, practices which, as conventional wisdom might have it, always and unproblematically reduced cartoons to children's entertainment, or which uniformly enforced censorship restrictions. Instead, the Hollywood cartoon from the classical era developed from, expressed, and was frequently controlled by a number of shifting and often contradictory discourses, about, for instance, sexuality, race, gender, class, leisure, and creativity. Other film practices within the movie industry were subject to the same or similar pressures, of course, but in this regard, a study of animation once again underlines its very special status.

Perceived as pure, unadulterated entertainment, cartoons were different from a number of Hollywood feature films—the gangster movies of the early 1930s, for instance, or the fallen woman films from the pre–World War II period, or the combat films produced during the war. These features dealt overtly with "real" social and political issues, and were discussed in the press, in publicity materials, and in social science studies as important cultural artifacts. The cartoon, however (and also cartoon producers) seemed concerned only with amusing a mass audience in general and a very young one in particular. But largely because of the obviousness of this motivation, cartoons worked in a number of settings as instruments of social control.

The same seems to be true in the 1990s, an era when theatrical animation has made a comeback as respectable, even artistically pleasing entertainment for a wide audience. And just as in the heyday of Hollywood animation, a broad system of discourses, including but not limited to the films themselves, functions to create and also to limit the possibilities for understanding cartoons and cartoon producers. In 1991, for instance, the *New York Times* referred to Steven Spielberg as "The Man Who Would Be Walt." The article about the movie mogul analyzed "the businessman in the boy," and discussed his toughness in negotiations which, nonetheless, led to the development of "fiercely loyal talent." Whether or not Spielberg "would be" Walt, then, the *Times*, and much of the media, have decided to construct him as a modern-day Disney, and use much the same language as the popular journals of the 1940s when they discussed the period's most famous cartoon producer. Of even more interest, I think, is that in trying to explain the Spielberg phenomenon, the *Times* constructed a series of binary oppositions—businessman/boy and toughness/devotion, for instance—that apparently provide a well-rounded picture of the producer of *Who Framed Roger Rabbit?* Disappearing from this portrait, however, is any mention of the use Spielberg makes of cheap labor in the Third World, or the problematic images of "zoot suiters" in *Roger Rabbit*, or of Arabs in the "Indiana Jones" films, or of Spielberg's position not just as a creator of family classics, but as an operative within transnational capitalism.

The *Times*' invocation of Disney demonstrates that, more than a quarter-century after his death, he still serves as the patron saint of animation, and perhaps as a result his company works diligently to control the manner in which his films can be interpreted by modern audiences. The recent and limited video release of *Fantasia*, for instance, in both regular and special editions, sought to turn that film into an unquestionable work of art, one that only a fortunate few million would be able to possess. Similarly, Disney tried to create a modern classic with 1991's *Beauty and the Beast*. The company clearly wanted to remind the public of the Disney tradition of making modern fairy tales, but also to convince viewers that the *Snow White*–style gender stereotypes of *The Little Mermaid* were simply aberrations, compensated for by the proto-feminist Beauty. A number of compliant film critics agreed, pointing out that this heroine likes to read books. Employing the 1940s discourse about Disney as a genius of reconciling opposites, the *Washington Post* added that the film was "a near-masterpiece," one "that draws on the sublime traditions of the past while remaining completely in sync with the sensibility of its time." Similarly, the television networks celebrated the movie as a fairy tale in the best Disney tradition, but also as a technological marvel possible only in the modern age of computerized filmmaking.[5] Going along with the electronic and print media, and certifying the film's status as instant classic, the Motion Picture Academy of Arts and Sciences made *Beauty and the Beast* the first cartoon to be nominated for a "Best Picture" Oscar. Thus, a range of related institutions locked in the possible readings of Disney's film; its greatness could be proved in terms of technology, narrative, and even politics.

In something of a departure from the Disney strategy, however, rather than locking in meaning Spielberg has sought seemingly to confound it, to make interpretation impossible by providing too many possibilities. *Who Framed Roger Rabbit?* functions as the ideal postmodern pastiche of cartoon practice, as animated characters from all eras and all studios make appearances. How, then, can the viewer make any sense out of Disney's Donald Duck and Warner Bros.' Daffy Duck, or MGM's Tom and Jerry and Paramount's Betty Boop, or Gertie the Dinosaur, from the teens, and Droopy, from the 1950s, sharing the screen?

Around the same time that Spielberg celebrated Hollywood cartoon history in *Roger Rabbit*, Warner Bros. rereleased a number of short animated films from the 1940s, 1950s, and 1960s to be shown before feature films at the AMC Theatre chain. Rather than simply creating nostalgia for the classical era's construction of entertainment, however, the reissue strategy created a response about race just as heated as that which followed Disney's release of *The New Spirit*. One of the reissued cartoons, *Sahara Hare* (1955), with Bugs Bunny, contained all of the 1950s stereotypes of Arabs, an orientalist discourse that angered numerous viewers.

From this, we can see that cartoon production and such related practices as popular journalism may not always have worked so smoothly at enforcing certain forms of social control and limiting possibilities for interpretation. As additional evidence, I have tried to show how the audience for cartoons went to see them for a variety of reasons and in a variety of settings, and that the cartoons themselves presented spectators with a number of interpretive possibilities. Thus, as the producers of popular culture, Disney, Schlesinger, the Fleischer Brothers, or Spielberg could not prevent their films from being, in fact, popular. That is, they could not keep them from being products of social orders that are always unstable and that unavoidably must be marked by conflict rather than consensus. As a result, in trying to reconstitute the goals of censorship, or the overriding logic of the film bill, or the consensus-building strategy of the journals, or the will of the government to control the governed, it is important not to see spectators merely as passive victims. Instead, we must look at the possibility for, and the power of, their resistance.

NOTES

Preface

1 David Bordwell, Janet Staiger, and Kristin Thompson, *The Classical Hollywood Cinema: Film Style and Mode of Production to 1960* (New York: Columbia University Press, 1985).

2 Lea Jacobs, *The Wages of Sin: Censorship and the Fallen Woman Film, 1928–1942* (Madison: University of Wisconsin Press, 1991); George Custen, *Bio/Pics: How Hollywood Constructed Public History* (New Brunswick, N.J.: Rutgers University Press, 1992); Jane M. Gaines, *Contested Culture: The Image, the Voice, and the Law* (Chapel Hill: University of North Carolina Press, 1991).

Hollywood Animation and Social Control

1 I am employing the term "Hollywood" in the manner of Rick Altman in *The American Film Musical* (Bloomington and Indianapolis: Indiana University Press, 1987), 123. According to Altman, "While 'Hollywood narrative film' . . . appears to represent a historical or geographical category, its recognition as a category in actuality stems from the fact that films of this type build a recognizable semantics into an identifiable syntax." As a result, even a feature film made in the 1920s at Paramount's Astoria Studios on Long Island would be considered a Hollywood film. I would add that a Hollywood film not only was made according to certain standardized production practices, but also was distributed and exhibited according to set systems established by the major film corporations. In relation to cartoons, this means that, even though the Fleischer Brothers produced their films in New York and Florida, their cartoons, which were distributed by Paramount, can be considered examples of Hollywood animation.

2 Raymond Spottiswoode, "Children in Wonderland," *The Saturday Review*, 13 November 1948, 5.

3 Letter and enclosure from J. Edgar Hoover to Assistant to the President, Major

General Wilton B. Persons, 16 March 1959, from the "Referrals" section of the FBI file on Walter Elias Disney, released through Freedom of Information-Privacy Act request.

4 For an extended discussion of the defining characteristics of the classical period, see Bordwell, Staiger, and Thompson, *The Classical Hollywood Cinema*. Generally, this period was marked by the domination of just a few major film studios and also by conventions regarding production practice, style and content, and theatrical exhibition.

5 Leonard Maltin, *Of Mice and Magic: A History of American Animated Cartoons* (New York: New American Library, 1980).

6 Ted Sennett, *Warner Brothers Presents* (Secaucus, N.J.: Castle Books, Inc., 1971), Appendix II.

7 Pierre Bourdieu, *Distinction: A Social Critique of the Judgement of Taste*, trans. Richard Nice (Cambridge, Mass.: Harvard University Press, 1984), 1.

8 Stephen Kern, *The Culture of Time and Space: 1880–1918* (Cambridge, Mass.: Harvard University Press, 1983).

9 Richard Schickel, *The Disney Version: The Life, Times, Art, and Commerce of Walt Disney* (New York: Simon and Schuster, 1968).

10 Dana Polan, *Power and Paranoia: History, Narrative, and the American Cinema, 1940–1950* (New York: Columbia University Press, 1986), 17.

Studio Strategies

1 For a discussion of the attempts made in this era to "control" motion picture content, see Garth Jowett, *Film: The Democratic Art* (Boston: Little, Brown and Company, 1976), chapter 4, "The Initial Response," and chapter 5, "The Movies Censored."

2 The Production Code is reprinted in Jowett, *Film: The Democratic Art*, 468–472. The "Reasons Supporting Preamble of Code" can be found on 471–472: I have cited sections 3A and 3C.

3 Lea Jacobs, "The Censorship of *Blonde Venus*: Textual Analysis and Historical Methods," *Cinema Journal* 27 (Spring 1988):21–31; Leonard J. Leff, "The Breening of America," *PMLA* 106 (May 1991):432–445.

4 "Hollywood Censors Its Animated Cartoons," *Look* magazine, 17 January 1939, 17–20.

5 Herbert Blumer, *Movies and Conduct* (New York: The Macmillan Company, 1933); Herbert Blumer and Philip M. Hauser, *Movies, Delinquency, and Crime* (New York: The Macmillan Company, 1933); Edgar Dale, *Children's Attendance at Motion Pictures* (New York: The Macmillan Company, 1935); Wendell S. Dysinger and Christian A. Ruckmick, *The Emotional Responses of Children to the Motion Picture Situation* (New York: The Macmillan Company, 1935); Peter W. Holaday and George D. Stoddard, *Getting Ideas from the Movies* (New York: The Macmillan Company, 1933); Mark A. May and Frank K. Shuttleworth, *The Social Conduct and Attitudes of Movie Fans* (New York: The Macmillan Company, 1933); Ruth C. Peterson and L. L. Thurstone, *Motion Pictures and the Social Attitudes of Children* (New York: The Macmillan Company, 1933); Charles C. Peters, *Motion Pictures and Standards of Morality* (New York: The Macmillan Company, 1933).

6 Lea Jacobs, "Reformers and Spectators: The Film Education Movement in the Thirties," *Camera Obscura* 22 (1990):29–49.

7 Barbara Deming, "The Library of Congress Film Project: Exposition of a Method," *The Library of Congress Quarterly Journal of Current Acquisitions* 2, no. 1 (1944):3–4.

8 Margaret Farrand Thorp, *America at the Movies* (New Haven: Yale University Press, 1939), 118, 120.

9 *Mickey Mouse Movie Stories*, story and illustrations by staff of Walt Disney Studio, introduction by Maurice Sendak (New York: Harry N. Abrams, Inc., 1988).

10 Jackson Lears, "A Matter of Taste: Corporate Cultural Hegemony in a Mass-Consumption Society," in *Recasting America: Culture and Politics in the Age of Cold War*, ed. Lary May (Chicago: University of Chicago Press, 1989), 39.

11 Polan, *Power and Paranoia*, 102.

12 Mickey Mouse, for instance, first appeared in 1928, Goofy (as Dippy Dawg) in 1931, and Donald Duck in 1934.

13 Michel Foucault, *The History of Sexuality*, Vol. 1, *An Introduction*, trans. Robert Hurley (New York: Pantheon Books, 1978), 6.

14 Malek Alloula, *The Colonial Harem*, trans. Myrna Godzick and Wlad Godzich (Minneapolis: University of Minnesota Press, 1986), 35.

15 Laura Mulvey, "Visual Pleasure and Narrative Cinema," *Screen* 16 (Autumn 1975):6–18.

16 Gail Minault and Hanna Papanek, eds., *Separate Worlds: Studies of Purdah in South Asia* (Delhi: Chanakya Publishers, 1982), 5.

17 "Boy Meets Facts," *Time* magazine, 21 July 1941, 73.

18 Joe Adamson, *Tex Avery: King of Cartoons* (New York: Da Capo Press, Inc., 1975), 182.

19 Richard Dyer, *Heavenly Bodies: Film Stars and Society* (New York: St. Martin's Press, 1986), 19.

20 Blumer, *Movies and Conduct*, 54.

21 Ibid., 58.

22 Showing the complexity of interpreting the film companies' adjustments to censorship, Heather Hendershot (letter to author, 8 February 1992) provides an alternate reading to the post-1934 "Betty Boop" credit sequence. Rather than seeing Betty as silhouetted against the window, Hendershot claims the credit shows Betty through a blind, thereby turning the spectator into a voyeur. Hence, rather than diminishing the sexual implications, these credits simply change the manner of presentation.

23 Annette Kuhn, *Cinema, Censorship, and Sexuality, 1909–1925* (London and New York: Routledge, 1988), 97.

24 For my analysis in this section I am indebted to Mark Anderson and his unpublished work on the comic body in silent films.

25 Maltin, *Of Mice and Magic*, 102.

26 For an examination of the ideology of Shirley Temple films, see Charles Eckert, "Shirley Temple and the House of Rockefeller," in *American Media and Mass Culture: Left Perspectives*, ed. Donald Lazere (Berkeley and Los Angeles: University of California Press, 1987), 164–177.

27 Dyer, *Heavenly Bodies*, 138.

28 The staff conducting the Library of Congress analyses was a small one, consisting of five people in 1944. Of the seven people who served as analysts between 1942

and 1944, five were women. See Deming, "The Library of Congress Film Project," 3.

29 For an analysis of this Production Code practice regarding "quality" films, see Leff, "The Breening of America."

30 Donald Crafton, "Walt Disney's *Peter Pan*: Woman Trouble on the Island," in *Storytelling in Animation: The Art of the Animated Image*, Vol. 2, ed. John Canemaker (Los Angeles: American Film Institute, 1988), 123–146.

Reading the Film Bill

1 *Washington Post*, 17 October 1931, p. 14.

2 Jack Alicoate, ed., *The 1937 Film Daily Yearbook of Motion Pictures* (N.p.: The Film Daily, 1937), 910.

3 Douglas Gomery explains this booking practice in *The Hollywood Studio System* (New York: St. Martin's Press, 1986), 18–19.

4 Federal Writers' Project, Works Progress Administration, *Washington: City and Capital*, "American Guide Series" (Washington, D.C.: U.S. Government Printing Office, 1937). For information on theater location, see page xx. For automobile information, page 7. The Guide adds that Detroit was second, with 1 car for every 5.07 residents, Baltimore third at 1 to 8.07, and New York fourth with 1 to 8.09.

5 Ibid., xx.

6 For population information, see ibid., 5–11.

7 Ibid., 6, 33.

8 For information regarding the clientele for English music halls, see Penny Summerfield, "Patriotism and Empire: Music-Hall Entertainment 1870–1914," in *Imperialism and Popular Culture*, ed. John M. MacKenzie (Manchester: Manchester University Press, 1986), 17–48. For vaudeville and cabaret, see Kern, *The Culture of Time and Space*, 200.

9 *Washington Post*, 31 October 1931, 14.

10 Ibid., 9 March 1934, 6.

11 Ibid., 2 May 1936, 16.

12 Ibid., 9 May 1936, 16.

13 Ibid., 31 October 1936, 18.

14 Ibid., 3 October 1931, 12.

15 Ibid., 2 December 1939, 18.

16 Ibid., 27 March 1937, 16.

17 Ibid., 9 March 1934, 6.

18 Maltin, *Of Mice and Magic*, 200.

19 For Agee's discussion of early movie theaters, see "Comedy's Greatest Era," *Life* magazine, 3 September 1949, reprinted in *Agee On Film*, Vol. 1 (New York: Grosset and Dunlap, 1969), 2–19.

20 Miriam Hansen, "Early Cinema: Whose Public Sphere?" in *Early Cinema: Space, Frame, Narrative*, ed. Thomas Elsaesser (London: British Film Institute, 1990), 228–246.

21 Gomery, *The Hollywood Studio System*, 14.

22 *Washington Post*, 6 August 1932, 14.

23 Ibid., 21 May 1932, 12.

24 Ibid., 28 August 1937, 22.

25 Ibid., 17 October 1931, 14; ibid., 6 August 1932, 14.

26 Ibid., 24 May 1936. The grid of "Film Bills at Residential Theaters This Week" showed that at the Milo in Rockville, Maryland, *Captain January* played with a cartoon and news, while the Apollo, in Washington, played the film with a "Mickey Mouse." The Central, the Cameo, and the Jesse showed the film with various live-action comic and educational shorts and newsreels. Several other theaters showing the film did not advertise their supplemental bills.

27 Ibid., 28 May 1932, 12.

28 Ibid., 6 March 1937, 12.

29 See, for instance, *Thirteen Women* (1931), starring Myrna Loy and Irene Dunne, or *High Flyers* (1938), a Wheeler and Woolsey comedy in which Margaret Dumont studies crystal-gazing.

30 For a discussion of Disney's place in world commerce, see Ariel Dorfman and Armand Mattelart, *How To Read Donald Duck: Imperialist Ideology in the Disney Comic*, trans. David Kunzle (New York: International General, 1975).

31 *Washington Post*, 17 October 1931, 14.

32 Ibid., 24 October 1931, 12.

33 Ibid., 7 August 1937, 12.

34 Ibid., 20 August 1938, 8.

35 Ibid., 23 December 1939, 6.

36 Ibid., 25 June 1942, 8.

37 For a discussion of this governmental concern, see Jowett, *Film: The Democratic Art*, 293–306.

The Disappearance of Dissent

1 For a description of this model of transmitting belief systems, see Martin Barker, *Comics: Ideology, Power, and the Critics* (Manchester: Manchester University Press, 1989), 4.

2 Richard Polenberg, *One Nation Divisible: Class, Race, and Ethnicity in the United States Since 1938* (New York: The Viking Press, 1980), 39.

3 Ibid., 69.

4 Charles C. Moskos, Jr., *The American Enlisted Man: The Rank and File in Today's Military* (New York: Russell Sage Foundation, 1970).

5 John Laffin, *Americans in Battle* (New York: Crown Publishers, Inc., 1973), 134.

6 William J. Blakefield, *Documentary Film Classics*, 2d ed. (Washington, D.C.: U.S. Government Printing Office, n.d.), 23.

7 Polan, *Power and Paranoia*, 55.

8 Moskos, *The American Enlisted Man*, 167.

9 Victor Hicken, *The American Fighting Man* (New York: The Macmillan Company, 1969), 147, 367.

10 Moskos, *The American Enlisted Man*, 119, 214.

11 Hicken, *The American Fighting Man*, 369.

12 Moskos, *The American Enlisted Man*, 202, 204.

13 Ibid., 70–71, 76.

14 Ibid., 146.

15 George Lipsitz, *Class and Culture in Cold War America: "A Rainbow at Midnight"* (New York: Praeger Publishers, 1981), 3.

16 Jacques Ellul, *Propaganda: The Formation of Men's Attitudes*, trans. Konrad Kellen and Jean Lerner (New York: Alfred A. Knopf, 1965), xi.

17 David Culbert, "*Why We Fight*: Social Engineering for a Democratic Society at War," in *Film and Radio Propaganda in World War II*, ed. K.R.M. Short (Knoxville: University of Tennessee Press, 1983), 173. In this essay, Culbert provides valuable information about the *Why We Fight* series.

18 Ellis Freeman, *Conquering the Man in the Street: A Psychological Analysis of Propaganda in War, Fascism, and Politics* (New York: Vanguard Press, 1940), 10, 28.

19 Ibid., 28.

20 Blakefield, *Documentary Film Classics*, 23.

21 Polan, *Power and Paranoia*, 163.

22 For an analysis of the male gaze in the cinema, see Mulvey, "Visual Pleasure and Narrative Cinema."

23 For an interesting discussion of this shift, see Alan Brinkley, "Prosperity, Depression, and War, 1920–1945," in *The New American History*, ed. Eric Fonor (Philadelphia: Temple University Press, 1990), 119–141.

24 Studies cited in Hicken, *The American Fighting Man*, 196.

25 Laffin writes that approximately 50 percent of all draftees were refused induction during the years 1941–1943 (*Americans in Battle*, 129).

26 Lipsitz, *Class and Culture in Cold War America*, 3.

27 Ibid., 8.

28 For this information about *The Home Front*, I have depended upon Jerry Beck and Will Friedwald, *Looney Tunes and Merrie Melodies: A Complete Illustrated Guide to the Warner Bros. Cartoons* (New York: Henry Holt and Company, 1989), 380.

29 Polan discusses this narrative system in *Power and Paranoia*, 55.

30 Ibid., 43.

The Popular Press Views Cartoons

1 *Time*: "Walt and the Professors," in the "Arts" column, 8 June 1942, 58–60; "Air: Sascha's Show," in the "Army & Navy" column, 12 July 1943, 63–64; "Mighty Mouse," in the "Merchandising" column, 25 October 1948, 96–98.

2 "The Dimensions of Disney," *The Saturday Review*, 6 June 1942, 5.

3 Ibid.

4 *Time*, 3 December 1945, 98.

5 *The Saturday Review*, 6 February 1943, 19.

6 Ibid.

7 "Mr. Disney's Caballeros," ibid., 24 February 1945, 22.

8 Ibid.

9 Ibid., 24.

10 *Time*, 7 June 1948, 98.

11 Ibid.

12 "Recessional," *The Saturday Review*, 3 June 1950, 30.

13 I discuss this early interaction of television and the film corporations in "Motion Pictures and Television, 1930–1945: A Pre-History of the Relations between the Two Media," *Journal of the University Film and Video Association* 34 (Summer 1982), 3–8.

14 Douglas Gomery, "Disney's Business History: A Reinterpretation," unpublished manuscript, 3–4.

15 See, for example, "The Pied Piper of Video," by Robert Lewis Shayon, *The Saturday Review*, 25 November 1950, 9–11, 49–51: "As for television's effects on children's emotional well-being and taste . . . TV is merely the newest battleground—admittedly the most crucial because it combines the appeals of movies and radio, is easily accessible to the child, and it is difficult to control his viewing of it" (10).

16 "Mr. Disney's Caballeros," *The Saturday Review*, 24.

17 "Recessional," ibid., 30.

18 *Time*, 18 November 1946, 101.

19 Ibid., 20 October 1947, 101.

20 Ibid., 20 February 1950, 89.

21 "Recessional," *The Saturday Review*, 30.

22 *Time*, 20 February 1950, 90.

23 "The Dimensions of Disney," *The Saturday Review*, 5.

24 *Time*, 27 October 1941, 98, 100.

25 Ibid., 9 March 1942, 82–83.

26 Ibid., 6 May 1946, 98–99.

27 Ibid., 18 November 1946, 101.

28 "Recessional," *The Saturday Review*, 28.

29 Ibid., 29.

30 "How Disney Combines Living Actors with His Cartoon Characters," *Popular Science*, September 1944, 107.

31 "Six-Inch Wax Dolls Are New Stars In Filmland," ibid., May 1946, 110.

32. "Ingenious Models Bring John Henry Folk Saga to Movie Life," ibid., December 1946, 113.

33 *Time*, 19 February 1945, 92.

34 Ibid., 6 May 1946, 101.

35 "Mr. Disney's Caballeros," *The Saturday Review*, 24.

36 "Recessional," ibid., 30.

37 *Time*, 19 February 1945, 92.

38 *The Saturday Review*, 24 February 1945, 23.

39 The review of *Make Mine Music* appears in *Time*, 6 May 1946, 98, 101; the Gallup Poll results are carried in "Boy Meets Facts," *Time*, 21 July 1941, 73.

40 "Merchandising," *Time*, 25 October 1948, 96 and 98.

41 "The Celluloid Civilization," *The Saturday Review*, 14 October 1950, 44.

42. Ibid., 45.

43 "Walt Disney Builds Half-Pint History," *Popular Science*, February 1953, 118–119.

44 "Walt and the Professors," *Time*, 8 June 1942, 60.

45 "Mickey Icarus," *The Saturday Review*, 4 September 1943, 18.

46 Leonard Mosley, *Disney's World* (New York: Stein and Day, 1985), 185–197.

47 Theodore Peterson, *Magazines in the Twentieth Century* (Urbana and Chicago: University of Illinois Press, 1964), 398.

48 George Wolseley, *Magazine World* (New York: Prentice-Hall, Inc., 1951), 56–69.

49 *Commonweal*, 8 August 1941, 377.

50 Jack Kinney, *Walt Disney and Assorted Other Characters: An Unauthorized Account of the Early Years at Disney's* (New York: Harmony Books, 1988), 137.

51 Peterson, *Magazines in the Twentieth Century*, 31.

52 Ibid., 239, 330.

53 *Business Week*, 16 March 1940, 44.

54 "Disney Developments," ibid., 14 June 1941, 46.

55 Ibid., 10 February 1945, 72.

56 Ibid., 72, 76.

57 Kinney, *Walt Disney and Assorted Other Characters*, 138.

58 *Business Week*, 10 February 1945, 76.

59 Kinney has written that, after the strike, "the studio was never the same . . . Walt cut off all privileges, and Disney's became a very hard-nosed place." See *Walt Disney and Assorted Other Characters*, 139. In his 1947 testimony before the House Committee on Un-American Activities, Disney insisted that the 1941 strike had been instigated by Communists. See "The Testimony of Walter E. Disney Before the House Committee on Un-American Activities," in *The American Animated Cartoon: A Critical Anthology*, ed. Gerald Peary and Danny Peary (New York: E. P. Dutton, 1980), 92–98.

60 Wolseley, *Magazine World*, 57.

61 Peterson, *Magazines in the Twentieth Century*, 210.

62 Ibid.

63 "The Dimensions of Disney," *The Saturday Review*, 5.

64 "Mickey Icarus," ibid., 19.

65 The photograph appears on page 19 of *The Saturday Review*, 4 September 1943.

66 *Atlantic Monthly*, December 1940, 692.

67 James Playsted Wood, *Magazines in the United States: Their Social and Economic Influence* (New York: The Ronald Press Company, 1949), 188.

68 Wolseley, *Magazine World*, 34.

69 *Time*, 9 February 1942, 36.

70 Ibid., 25 January 1943, 86.

71 Kinney, *Walt Disney and Assorted Other Characters*, 139.

72 *Time*, 3 December 1945, 98.

73 "Show Business," *Time*, 12 August 1946, 86.

74 Peterson, *Magazines in the Twentieth Century*, 371–372.

75 *Popular Science*, September 1944, 108.

76 Ibid.

77 Ibid., 110.

78 Ibid.

Disney Diplomacy

1 Some examples of these studio histories include: Tino Balio, *United Artists: The Company Built by the Stars* (Madison: University of Wisconsin Press, 1975); Gomery, *The Hollywood Studio System*; Thomas Schatz, *The Genius of the System: Hollywood Filmmaking in the Studio Era* (New York: Pantheon Books, 1988); Neal Gabler, *An Empire of Their Own: How the Jews Invented Hollywood* (New York: Crown Publishers, Inc., 1988).

2 The Disney documents from the State and Treasury Departments are housed in the National Archives in Washington, D.C. The State Department papers are in a file labeled "Adult Education, 1940–44." Treasury Department records are in a file called "Disney, Walt, Productions (Donald Duck Film *The New Spirit*)." I received the FBI documents through my Freedom of Information-Privacy Act (FOIPA) request number 316,469. These documents are contained in four separate files, marked "Miscellaneous Cross-Reference," "HQ 94–4–4667," "9–33728," and "Los Angeles 80–294." The first three files have as their subject "Walter Elias Disney," the last, "Walt Disney." Documents from all of the files are difficult to cite, because pages are not numbered and they are not arranged chronologically. Within this chapter, I will refer to these documents by date.

3 Andre Gunder Frank, *Capitalism and Underdevelopment in Latin America* (New York: Monthly Review Press, 1967), 8.

4 Richard Shale provides an excellent narrative of Disney's tour in *Donald Duck Joins Up: The Walt Disney Studio during World War II* (Ann Arbor, Mich.: UMI Research Press, 1976), 41–49.

5 David Cook, *A History of Narrative Film*, 2d ed. (New York: W. W. Norton & Co., 1990), 824.

6 Juan Eugencio Corradi, "Argentina," in *Latin America: The Struggle with Dependency and Beyond*, ed. Ronald H. Chilcote and Joel C. Edelstein (Cambridge, Mass.: Shenkman Publishing Co., Inc., 1974), 358.

7 J.C.M. Ogelsby, "Who Are We? The Search for a National Identity in Argentina, Australia, and Canada, 1870–1950," in *Argentina, Australia, and Canada: Studies in Comparative Development, 1870–1965*, ed. D.C.M. Platt and Guido Di Tella (London: The Macmillan Press Ltd., 1985), 115–116.

8 Theotonio Dos Santos, "Brazil: The Origins of a Crisis," in *Latin America: The Struggle with Dependency and Beyond*, ed. Chilcote and Edelstein, 445.

9 Carlos F. Diaz Alejandro, "Latin America in the 1930's," in *Latin America in the 1930's: The Role of the Periphery in World Crisis*, ed. Rosemary Thorp (London: The Macmillan Press Ltd., 1984), 17–18, 33, 37.

10 Mosley, *Disney's World*, 195–196: Kinney, *Walt Disney and Assorted Other Characters*, 139.

11 Corradi, "Argentina," 358.

12 Felix J. Wall, *Argentine Riddle* (New York: The John Day Company, 1944), 173: Corradi, "Argentina," 359.

13 Corradi, "Argentina," 359.

14 Andre Gunder Frank, *Latin America: Underdevelopment or Revolution* (New York: Modern Reader, 1969), 152.

15 Marcello De Paiva Abreu, "Argentina and Brazil during the 1930's: The Impact of British and American International Economic Policies," in *Latin America in the 1930's*, ed. Thorp, 149, 151.

16 John T. Elliff, "The Scope and Basis of FBI Data Collection," in *Investigating the FBI*, ed. Stephen Gillers and Pat Watters (Garden City, N.Y.: Doubleday and Co., Inc., 1973), 258.

17 Richard E. Morgan, *Domestic Intelligence: Monitoring Dissent in America* (Austin: University of Texas Press, 1980), 40.

18 Ibid., 38.

19 Ibid., 34.

20 Thomas I. Emerson, "The FBI as a Political Police," in *Investigating the FBI*, ed. Watters and Gillers, 242.

21 Nancy Lynn Schwartz, *The Hollywood Writers Wars* (New York: Alfred A. Knopf, 1982), 257.

22 For a production history of *The New Spirit* and also of Disney's next Treasury film, *The Spirit of '43*, see Shale, *Donald Duck Joins Up*, 27–35.

23 U.S. Congress, *Congressional Record: Proceedings and Debates of the 77th Congress, Second Session* (Washington, D.C.: U.S. Government Printing Office), 9 February 1942, 1160–1162.

24 For a discussion of the relationship between Congress and the movie studios both before and after Pearl Harbor, see Jowett, *Film: The Democratic Art*, chapter 12, "Hollywood Goes to War, 1939–1945," 295–332.

25 See Cathy Klaprat, "The Star as Market Strategy: Bette Davis in Another Light," in *The American Film Industry*, ed. Tino Balio (Madison: University of Wisconsin Press, 1985), 351–376. Also, Maria La Place, "Bette Davis and the Ideal of Consumption," *Wide Angle* 6 (Fall 1984):34–44.

26 Bernard F. Dick, *The Star-Spangled Screen: The American World War II Film* (Lexington: University Press of Kentucky, 1985). Dick discusses the internment decision as well as "anti-Nippon" films on page 236.

Afterword

1 Carolyn Marvin, *When Old Technologies Were New: Thinking about Electric Communication in the Late Nineteenth Century* (New York: Oxford University Press, 1988), 233.

2 William Grimes, "Cartoons with a Political Conscience," *New York Times*, 7 February 1992, C3.

3 Charles Solomon, "When American Cartoons Went Off to War," *Los Angeles Times*, 7 February 1992, F12.

4 John Fiske, *Understanding Popular Culture* (Winchester, Mass.: Unwin Hyman, 1989), 159.

5 Hal Hinson, "Beautiful *Beast*: Disney's Fairest Fairy Tale," *Washington Post*, 22 November 1991, B6.

BIBLIOGRAPHY

Unpublished Sources

FBI documents related to Walt Disney, released through Freedom of Information-Privacy Act request number 316,469.

Motion picture copyright documents, held in the Library of Congress, Motion Picture Division, Washington, D.C.

State Department documents, "Adult Education, 1940–44," 810.42711, housed in National Archives, Washington, D.C.

Treasury Department documents, "Disney, Walt, Productions (Donald Duck Film *The New Spirit*)," held in National Archives, Washington, D.C.

Published Sources

Adamson, Joe. *Tex Avery: King of Cartoons.* New York: Da Capo Press, Inc., 1975.

Agee, James. *Agee on Film*, Vol. 1. New York: Grosset and Dunlap, 1969.

Alicoate, Jack, ed. *The 1937 Film Daily Yearbook of Motion Pictures.* N.p.: The Film Daily, 1937.

Alloula, Malek. *The Colonial Harem.* Translated by Myrna Godzick and Wlad Godzich. Minneapolis: University of Minnesota Press, 1986.

Altman, Rick. *The American Film Musical.* Bloomington and Indianapolis: Indiana University Press, 1987.

Balio, Tino. *United Artists: The Company Built by the Stars.* Madison: University of Wisconsin Press, 1975.

———, ed. *The American Film Industry.* Madison: University of Wisconsin Press, 1985.

Barker, Martin. *Comics: Ideology, Power, and the Critics.* Manchester: Manchester University Press, 1989.

Beck, Jerry, and Will Friedwald. *Looney Tunes and Merrie Melodies: A Complete Illustrated Guide to the Warner Bros. Cartoons*. New York: Henry Holt and Company, 1989.

Blakefield, William J. *Documentary Film Classics*. 2d ed. Washington, D.C.: U.S. Government Printing Office, n.d.

Blumer, Herbert. *Movies and Conduct*. New York: The Macmillan Company, 1933.

Blumer, Herbert, and Philip M. Hauser. *Movies, Delinquency, and Crime*. New York: The Macmillan Company, 1933.

Bordwell, David, Janet Staiger, and Kristin Thompson. *The Classical Hollywood Cinema: Film Style and Mode of Production to 1960*. New York: Columbia University Press, 1985.

Bourdieu, Pierre. *Distinction: A Social Critique of the Judgement of Taste*. Translated by Richard Nice. Cambridge, Mass.: Harvard University Press, 1984.

Callaway, Helen. *Gender, Culture, and Empire: European Women in Colonial Nigeria*. Urbana and Chicago: University of Illinois Press, 1987.

Chilcote, Ronald H., and Joel C. Edelstein. *Latin America: The Struggle with Dependency and Beyond*. Cambridge, Mass.: Shenkman Publishing Co., Inc., 1974.

Cook, David. *A History of Narrative Film*. 2d ed. New York: W. W. Norton & Co., 1990.

Crafton, Donald. "Walt Disney's *Peter Pan*: Woman Trouble on the Island." In *Storytelling in Animation: The Art of the Animated Image*, Vol. 2, ed. John Canemaker. Los Angeles: American Film Institute, 1988.

Custen, George. *Bio/Pics: How Hollywood Constructed Public History*. New Brunswick, N.J.: Rutgers University Press, 1992.

Dale, Edgar. *Children's Attendance at Motion Pictures*. New York: The Macmillan Company, 1935.

de Cordova, Richard. *Picture Personalities: The Emergence of the Star System in America*. Urbana and Chicago: University of Illinois Press, 1990.

Deming, Barbara. "The Library of Congress Film Project: Exposition of a Method." *The Library of Congress Quarterly Journal of Current Acquisitions* 2, no. 1 (1944):3–36.

Dick, Bernard F. *The Star-Spangled Screen: The American World War II Film*. Lexington: University Press of Kentucky, 1985.

Dorfman, Ariel, and Armand Mattelart. *How to Read Donald Duck: Imperialist Ideology in the Disney Comic*. Translated by David Kunzle. New York: International General, 1975.

Dyer, Richard. *Heavenly Bodies: Film Stars and Society*. New York: St. Martin's Press, 1986.

Dysinger, Wendell S., and Christian A. Ruckmick. *The Emotional Responses of Children to the Motion Picture Situation*. New York: The Macmillan Company, 1935.

Eckert, Charles. "Shirley Temple and the House of Rockefeller." In *American Media and Mass Culture: Left Perspectives*, ed. Donald Lazere. Berkeley and Los Angeles: University of California Press, 1987.

Ellul, Jacques. *Propaganda: The Formation of Men's Attitudes*. Translated by Konrad Kellen and Jean Lerner. New York: Alfred A. Knopf, 1965.

Federal Writers' Project, Works Progress Administration. *Washington: City and Capital*. "American Guide Series." Washington, D.C.: U.S. Government Printing Office, 1937.

Fiske, John. *Understanding Popular Culture*. Winchester, Mass.: Unwin Hyman, 1989.

Fonor, Eric, ed. *The New American History*. Philadelphia: Temple University Press, 1990.

Frank, Andre Gunder. *Capitalism and Underdevelopment in Latin America*. New York: Monthly Review Press, 1967.

———. *Latin America: Underdevelopment or Revolution*. New York: Modern Reader, 1969.

Freeman, Ellis. *Conquering the Man in the Street: A Psychological Analysis of Propaganda in War, Fascism, and Politics*. New York: Vanguard Press, 1940.

Gabler, Neal. *An Empire of Their Own: How the Jews Invented Hollywood*. New York: Crown Publishers, Inc., 1988.

Gaines, Jane M. *Contested Culture: The Image, the Voice, and the Law*. Chapel Hill: University of North Carolina Press, 1991.

Gillers, Stephen, and Pat Watters, eds. *Investigating the FBI*. Garden City, N.Y.: Doubleday and Co., Inc., 1973.

Goldberg, David Theo, ed. *Anatomy of Racism*. Minneapolis: University of Minnesota Press, 1990.

Gomery, Douglas. *The Hollywood Studio System*. New York: St. Martin's Press, 1986.

Hansen, Miriam. "Early Cinema: Whose Public Sphere?" In *Early Cinema: Space, Frame, Narrative*, ed. Thomas Elsaesser. London: British Film Institute, 1990.

Hicken, Victor. *The American Fighting Man*. New York: The Macmillan Company, 1969.

Holaday, Peter W., and George D. Stoddard. *Getting Ideas from the Movies*. New York: The Macmillan Company, 1933.

Jacobs, Lea. "The Censorship of *Blonde Venus*: Textual Analysis and Historical Methods." *Cinema Journal* 27 (Spring 1988):21–31.

———. "Reformers and Spectators: The Film Education Movement in the Thirties." *Camera Obscura* 22 (January 1990):29–49.

———. *The Wages of Sin: Censorship and the Fallen Woman Film, 1928–1942*. Madison: University of Wisconsin Press, 1991.

Jowett, Garth. *Film: The Democratic Art*. Boston: Little, Brown and Company, 1976.

Kern, Stephen. *The Culture of Time and Space: 1880–1918*. Cambridge, Mass.: Harvard University Press, 1983.

Kinney, Jack. *Walt Disney and Assorted Other Characters: An Unauthorized Account of the Early Years at Disney's*. New York: Harmony Books, 1988.

Kuhn, Annette. *Cinema, Censorship and Sexuality, 1909–1925*. London and New York: Routledge, 1988.

Laffin, John. *Americans in Battle*. New York: Crown Publishers, Inc., 1973.

La Place, Maria. "Bette Davis and the Ideal of Consumption." *Wide Angle* 6 (Fall 1984):34–44.

Leff, Leonard J. "The Breening of America." *PMLA* 106 (May 1991):432–445.

Lipsitz, George. *Class and Culture in Cold War America: "A Rainbow at Midnight."* New York: Praeger Publishers, 1981.

McElvaine, Robert S. *The Great Depression: America 1929–1941*. New York: Times Books, 1984.

MacKenzie, John M., ed. *Imperialism and Popular Culture*. Manchester: Manchester University Press, 1986.

Maland, Charles J. *Chaplin and American Culture*. Princeton, N.J.: Princeton University Press, 1989.

Maltin, Leonard. *Of Mice and Magic: A History of America Animated Cartoons*. New York: New American Library, 1980.

Marchetti, Gina. "Action Adventure as Ideology." In *Cultural Politics in Contemporary America*, ed. Ian Angus and Sut Jhally. New York and London: Routledge, 1989.

Martin, Olga J. *Hollywood's Movie Commandments: A Handbook for Motion Picture Writers and Reviewers*. New York: H. W. Wilson Company, 1937.

Marvyn, Carolyn. *When Old Technologies Were New: Thinking about Electric Communication in the Late Nineteenth Century*. New York: Oxford University Press, 1988.

Mattelart, Armand, and Seth Siegelaub, eds. *Communication and Class Struggle*, Vol. 1. New York: International General, 1979.

May, Elaine Tyler. *Homeward Bound: American Families in the Cold War Era*. New York: Basic Books, Inc., 1988.

May, Lary, ed. *Recasting America: Culture and Politics in the Age of Cold War*. Chicago: University of Chicago Press, 1989.

May, Mark A., and Frank K. Shuttleworth. *The Social Conduct and Attitudes of Movie Fans*. New York: The Macmillan Company, 1933.

Minault, Gail, and Hanna Papanek, eds. *Separate Worlds: Studies of Purdah in South Asia*. Delhi: Chanakya Publishers, 1982.

Morgan, Richard E. *Domestic Intelligence: Monitoring Dissent in America*. Austin: University of Texas Press, 1980.

Mosley, Leonard. *Disney's World*. New York: Stein and Day, 1985.

Moskos, Charles C., Jr. *The American Enlisted Man: The Rank and File in Today's Military*. New York: Russell Sage Foundation, 1970.

Mulvey, Laura. "Visual Pleasure and Narrative Cinema." *Screen* 16 (Autumn 1975):6–18.

Peary, Gerald, and Danny Peary, eds. *The American Animated Cartoon: A Critical Anthology*. New York: E. P. Dutton, 1980.

Peters, Charles C. *Motion Pictures and Standards of Morality*. New York: The Macmillan Company, 1933.

Peterson, Ruth C., and L. L. Thurstone. *Motion Pictures and the Social Attitudes of Children*. New York: The Macmillan Company, 1933.

Peterson, Theodore. *Magazines in the Twentieth Century*. Urbana and Chicago: University of Illinois Press, 1964.

D.C.M. Platt, and Guido Di Tella. *Argentina, Australia, and Canada: Studies in Comparative Development, 1870–1965*. London: The Macmillan Press Ltd., 1985.

Polan, Dana. "Daffy Duck and Bertolt Brecht: Toward a Politics of Self-Reflexive Cinema?" In *American Media and Mass Culture: Left Perspectives*, ed. Donald Lazere. Berkeley and Los Angeles: University of California Press, 1987.

———. *Power and Paranoia: History, Narrative, and the American Cinema, 1940–1950*. New York: Columbia University Press, 1986.

Polenberg, Richard. *One Nation Divisible: Class, Race, and Ethnicity in the United States Since 1938*. New York: The Viking Press, 1980.

Said, Edward. *Orientalism*. New York: Pantheon Books, 1978.

Schatz, Thomas. *The Genius of the System: Hollywood Filmmaking in the Studio Era*. New York: Pantheon Books, 1988.

Schickel, Richard. *The Disney Version: The Life, Times, Art, and Commerce of Walt Disney*. New York: Simon and Schuster, 1968.

Schwartz, Nancy Lynn. *The Hollywood Writers Wars*. New York: Alfred A. Knopf, 1982.

Shale, Richard. *Donald Duck Joins Up: The Walt Disney Studio during World War II*. Ann Arbor, Mich.: UMI Research Press, 1976.

Shohat, Ella. "Gender and the Culture of Empire: Toward a Feminist Ethnography of the Cinema." *Quarterly Review of Film and Video* 13, nos. 1–3 (1991):45–84.

Short, K.R.M., ed. *Film and Radio Propaganda in World War II*. Knoxville: University of Tennessee Press, 1983.

Staff of Walt Disney Studio. *Mickey Mouse Movie Stories*. New York: Harry N. Abrams, Inc., 1988.

Syzmanski, Albert. *Class Structure: A Critical Perspective*. New York: Praeger Publishers, 1983.

Thorp, Margaret Farrand. *America at the Movies*. New Haven: Yale University Press, 1939.

Thorp, Rosemary, ed. *Latin America in the 1930's: The Role of the Periphery in World Crisis*. London: The Macmillan Press Ltd., 1984.

Wall, Felix J. *Argentine Riddle*. New York: The John Day Company, 1944.

Wolseley, George. *Magazine World*. New York: Prentice-Hall, Inc., 1951.

Wood, James Playsted. *Magazines in the United States: Their Social and Economic Influence*. New York: The Ronald Press Company, 1949.

Zipes, Jack. *Breaking the Magic Spell: Radical Theories of Folk and Fairy Tales*. New York: Methuen, 1979.

INDEX

Note: Page numbers for illustrations are in boldface type. Cartoon characters have been alphabetized by their first names. Thus, José Carioca has been placed under J, not C.

Printed in the United States
1092800005B